I0488830

Multi-Queue

SBC Observation Congruence

Model

For Systems Architecture

-- The Structure-Behavior Coalescence Approach --

William S. Chao

Structure-Behavior Coalescence

$$\text{Systems Architecture} = \text{Systems Structure} + \text{Systems Behavior}$$

The two processes are said to be observation congruent if their observational behaviors are the same, but internal behaviors may differ widely.

CONTENTS

8

10

ABOUT THE AUTHOR

Dr. William S. Chao is the CEO & founder of SBC Architecture International®. SBC (Structure-Behavior Coalescence) architecture is a systems architecture which demands the integration of systems structure and systems behavior of a system. SBC architecture applies to hardware architecture, software architecture, enterprise architecture, knowledge architecture and thinking architecture. The core theme of SBC architecture is: Architecture = Structure + Behavior.

William S. Chao received his bachelor degree (1976) in telecommunication engineering and master degree (1981) in information engineering, both from the National Chiao-Tung University, Taiwan. From 1976 till 1983, he worked as an engineer at Chung-Hwa Telecommunication Company, Taiwan.

William S. Chao received his master degree (1985) in information science and Ph.D. degree (1988) in information science, both from the University of Alabama at Birmingham, USA. From 1988 till 1991, he worked as a computer scientist at GE Research and Development Center, Schenectady, New York, USA.

Dr. William S. Chao has been teaching at National Sun Yat-

Sen University, Taiwan since 1992 and now serves as the president of Association of Enterprise Architects, Taiwan Chapter. His research covers: systems architecture, hardware architecture, software architecture, enterprise architecture, knowledge architecture and thinking architecture.

PART I: WHAT IS PROCESS ALGEBRA?

Algebraic Approach to the Study of Concurrent Systems

Process algebras are a diverse family of related approaches to the study of concurrent systems. Their tools are algebraic languages for the high-level description of interactions, communications and synchronizations between a collection of independent agents or processes.

Process algebras also provide algebraic laws that allow process descriptions to be manipulated and analyzed, and permit formal reasoning about equivalences and observation congruence among processes.

Examples of Process Algebras

There are several leading algebraic approaches to modeling concurrent systems.

Communicating Sequential Processes (CSP) was first described in a 1978 paper by C. A. R. Hoare.

Arthur John Robin Gorell Milner introduced the Calculus of Communicating Systems (CCS) around 1980.

Algebra of Communicating Processes (ACP) was initially developed by Jan Bergstra and Jan Willem Klop in 1982.

Multi-Queue SBC Process Algebra

Single-queue SBC process algebra, multi-queue SBC process algebra and infinite-queue SBC process algebra are the three SBC process algebras.

Multi-queue SBC process algebra evolved from CCS (Calculus of Communicating Systems).

CCS is a general process algebra language for the study of concurrent systems. Unlike CCS, multi-queue SBC process algebra is only applicable to systems architecture.

PART II: MATHEMATICS OF MULTI-QUEUE SBC PROCESSES

Interaction

An interaction represents an indivisible and instantaneous handshake or communication between two agents. The caller agent (either external environment's actor or component) communicates with the callee agent (component) through the interaction.

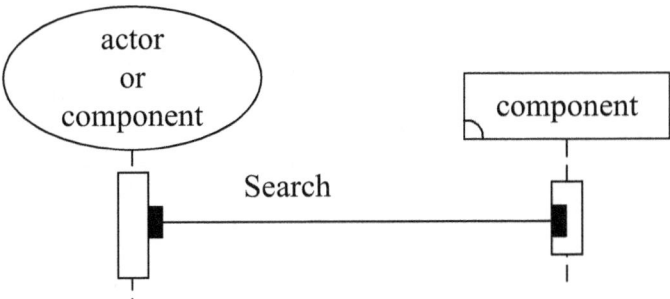

There are two ports, i.e., calling port or called port, of an interaction. The caller agent owns the "calling port" of the interaction.

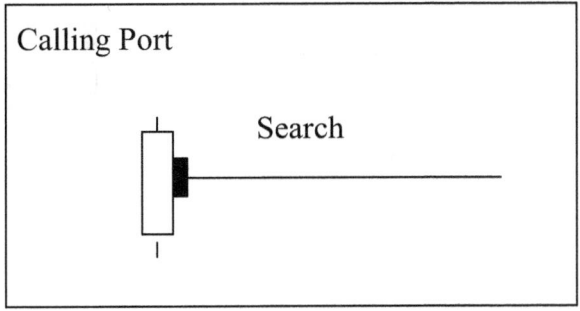

The caller agent together with the "calling port" is named the "calling action".

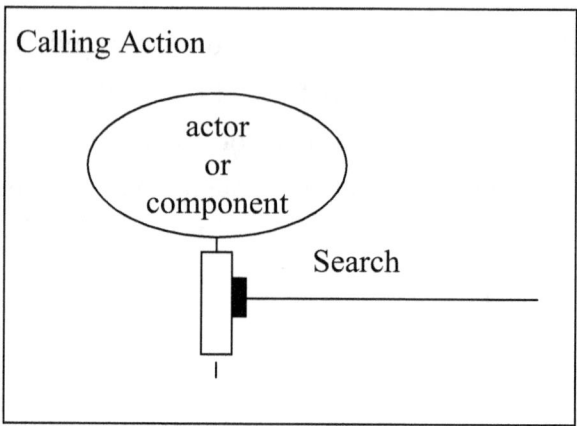

The callee agent owns the "called port" of the interaction.

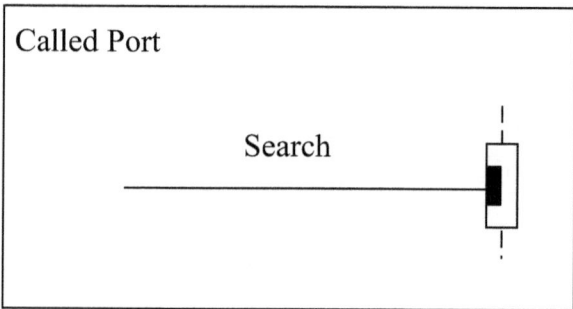

The callee agent together with the "called port" is named the "called action".

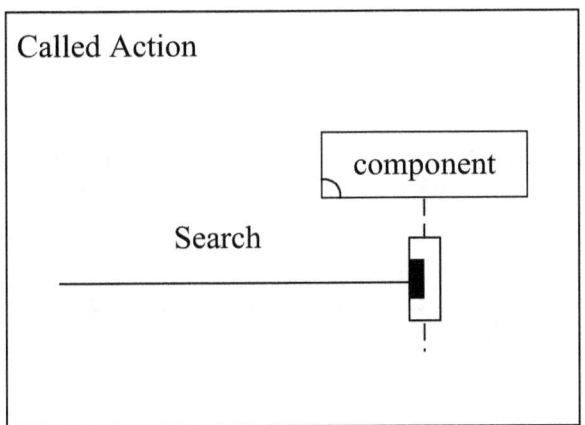

In order to simplify the interaction diagram, we will redraw it as follows.

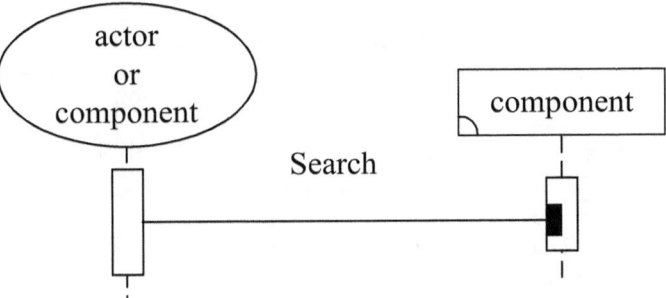

We use an internal interaction (i.e. λ) to represent their handshake or communication, if the caller agent and the callee agent are the same component.

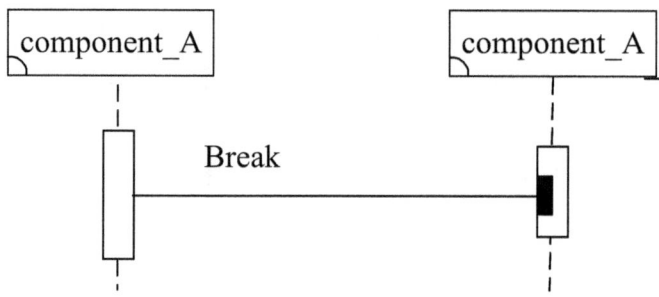

Also, we may redraw the internal interaction as follows.

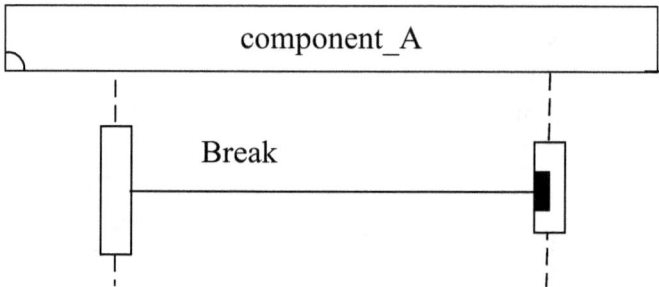

Sequentialization of Interactions

Sometimes interactions must be temporally ordered. For example, it might be desirable to specify algorithms such as: execute the "α" interaction first and then execute the "P" process later. Sequentialization of interactions can be used for such purposes.

Sequentialization of interactions, usually written as the $\alpha \bullet P$ process, indicates that it will perform the "α" interaction first and continue as the "P" process.

Parallel Composition of Processes

Parallel composition of two processes P and Q, usually written $P\|Q$, is the key primitive distinguishing the process algebras from sequential models of process executions.

Parallel composition allows the executions in P and Q to proceed simultaneously and independently.

Recursive Definition of a Process

The operators presented so far describe only finite interaction and are consequently insufficient for full computability, which includes non-terminating behavior. Recursion is the operator that allows finite descriptions of infinite behavior.

For example, **fix**($X=E$) can be understood as abbreviating the recursive definition of an infinite behavior denoted by the "X" process variable.

Conditional Definition of a Process

A process can be defined by a one-or-more-armed conditional expression.

For example, the process (**if** $cond_1$ **then** P_1)+(**if** $cond_2$ **then** P_2)...+(**if** $cond_j$ **then** P_j) will proceed as the process P_1 if the "$cond_1$" value is true, or proceed as the process P_2 if the "$cond_2$" value is true,..., or proceed as the process P_j if the "$cond_j$" value is true.

Null Process

Process algebras generally also include a null process, denoted as *STOP*, which has no interaction points. It is utterly inactive and its sole purpose is to act as the inductive anchor on top of which more interesting processes can be generated.

The process "*STOP•P*" (i.e. sequential composition of processes *STOP* and *P*) equals to the process "*STOP*".

$$STOP \bullet P \quad = \quad STOP$$

The process "*P‖STOP*" (i.e. parallel composition of processes *P* and *STOP*) equals to the process "*STOP‖P*" (i.e. parallel composition of processes *STOP* and *P*) which equals to the process "*P*".

$$P \parallel STOP \quad = \quad STOP \parallel P \quad = \quad P$$

PART III: THE STRUCTURE-BEHAVIOR COALESCENCE APPROACH

Structure-Behavior Coalescence Means to Integrate the Systems Structure and Systems Behavior

Systems structure and systems behavior are the two most prominent views of a system, integrating the systems structure and systems behavior is apparently the best way to achieve an integrated whole of a system.

If we are not able to integrate the systems structure and systems behavior, then there is no way that we are able to integrate the whole system.

Structure-behavior coalescence (SBC) provides an elegant way to integrate the systems structure and systems behavior of a system. In other words, SBC facilitates an integrated whole of a system.

Core Theme of Structure-Behavior Coalescence

The core theme of structure-behavior coalescence is: "Systems Architecture = Systems Structure + Systems Behavior."

Systems Structure X ╈ Systems Behavior X

One systems structure will draw forth one systems behavior. That is, the systems behavior is attached to or built on the systems structure in the SBC approach.

In other words, the systems behavior can not exist alone; it must be loaded on the systems structure just like a cargo is loaded on a ship.

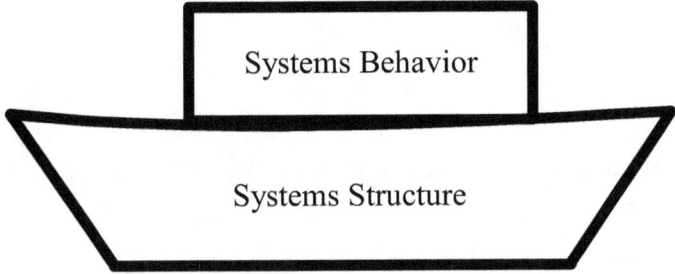

Interactions among Components and Actors to Draw Forth the Systems Behavior

In a system, if the components, and among them and the external environment's actors to interact (or handshake), these interactions will draw forth the systems behavior.

We conclude that "interaction" plays an important factor in integrating the systems structure and systems behavior for a system.

The overall behavior of a system consists of many individual behaviors. Each individual behavior represents an execution path. We use an interaction flow diagram (IFD) to demonstrate this individual behavior.

SBC View Model

In the SBC view model, dimension 1 stands for the evolution&motivation view which contains the strategy/version 1, strategy/version 2, strategy/version 3, strategy/version 4, strategy/version ∞ views; dimension 2 stands for the multi-level view which contains the concept, analysis, design and implementation views; dimension 3 stands for the systemic view which contains the structure and behavior views.

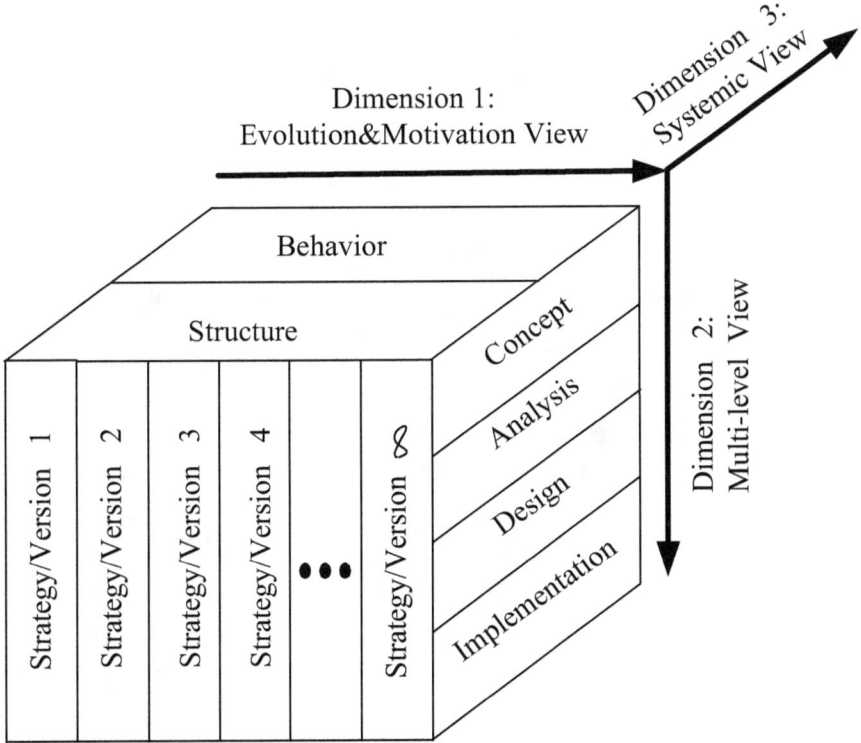

Collection of All Interaction Flow Diagrams Defines the Systemic View of Systems Architecture

The collection of all interaction flow diagrams defines the integration of systems structure and systems behavior of a system.

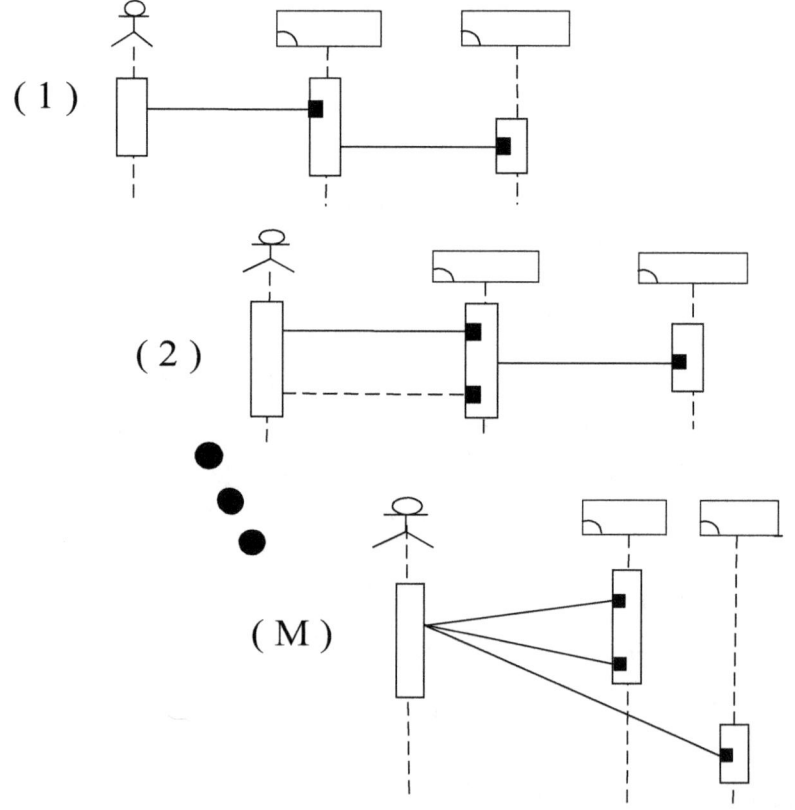

That is, the collection of all interaction flow diagrams defines the systemic view of systems architecture.

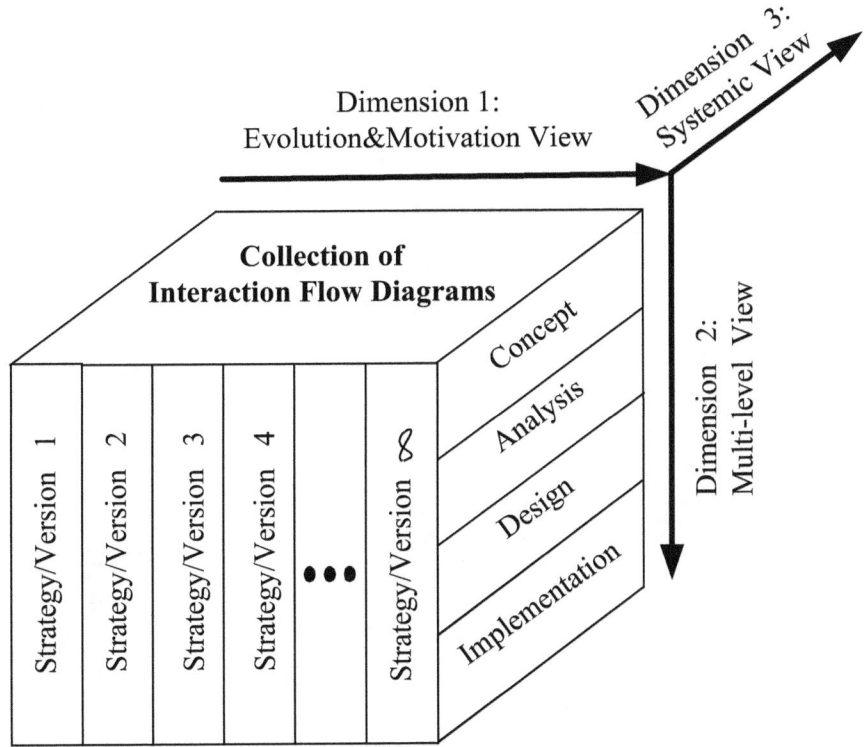

PART IV: LANGUAGE CONSTRUCTS OF MULTI-QUEUE SBC PROCESS ALGEBRA

Backus-Naur Form of Multi-Queue SBC Processes

The set of multi-queue SBC (i.e. Structure-Behavior Coalescence) processes for (the structural composition of) the systemic view of each level (i.e. Concept, Analysis, Design, Implementation) is defined by the following BNF grammar:

(1) <Structural Composition of Multi-Queue_SBC_Process_of_Each_Level's_Systemic_View>::=
    <Multi-Queue_SBC_Process_of_Each_Level's_Systemic_View>[*f*]
(2) <Multi-Queue_SBC_Process_of_Each_Level's_Systemic_View> ::= <Parallel_of_FixIFD>
(3) <Parallel_of_FixIFD> ::=   *STOP*
    | "**fix**("<Process_Variable>"=" <IFD><Process_Variable>")"
    "||" <Parallel_of_FixIFD>
(4) <IFD> ::= <Type_1_Expression> <Zero_Or_More_Expressions>
(5) <Zero_Or_More_Expressions> ::= " ● "
    | " ● " <Type_1_Or_2_Expression> <Zero_Or_More_Expressions>
(6) <Type_1_Or_2_Expression> ::= <Type_1_Expression>
    | <Type_2_Expression>
(7) <Type_1_Expression> ::= <Type_1_Interaction>
    | <Condition> <Type_1_Interaction>
    {"+" <Condition> <Type_1_Interaction>}
(8) <Type_2_Expression> ::= <Type_2_Interaction>
    | <Condition> <Type_2_Interaction>
    {"+" <Condition> <Type_2_Interaction>}
(9) <Type_1_Interaction> ::= <Actor> <Operation_Call_Or_Return>
    <Operation_Call_Or_Return_Formula> <Component>
(10) <Type_2_Interaction> ::= <Component> <Operation_Call_Or_Return>
    <Operation_Call_Or_Return_Formula> <Component>

A Component Interacting with another Component Defines the Type_2 Interaction

Rule 10 describes that a component interacting with another component defines the type_2 interaction.

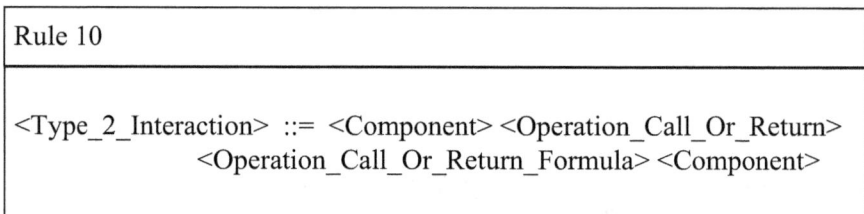

Rule 10

<Type_2_Interaction> ::= <Component> <Operation_Call_Or_Return>
<Operation_Call_Or_Return_Formula> <Component>

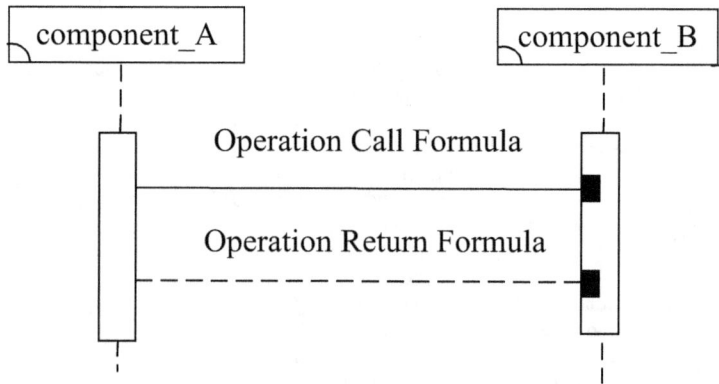

An Actor Interacting with a Component Defines the Type_1 Interaction

Rule 9 describes that an actor interacting with a component defines the type_1 interaction.

Rule 9
<Type_1_Interaction> ::= <Actor> <Operation_Call_Or_Return> <Operation_Call_Or_Return_Formula> <Component>

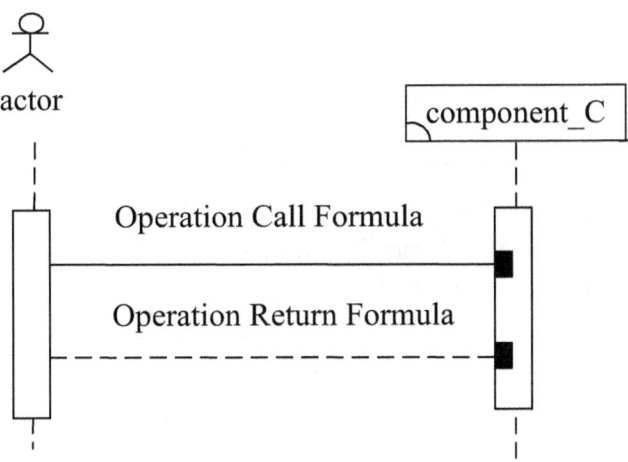

Type_2 Expression is either a Type_2 Interaction or a Conditional Type_2 Interaction

Rule 8 describes that the type_2 expression (i.e. Type_2_Expression) is either a type_2 interaction (i.e. Type_2_Interaction) or a conditional type_2 interaction (i.e. one-or-more-armed conditional expression of Type_2_Interaction).

Rule 8

<Type_2_Expression>   ::=   <Type_2_Interaction>
  | <Condition> <Type_2_Interaction>
     {"+" <Condition> <Type_2_Interaction>}

Type_1 Expression is either a Type_1 Interaction or a Conditional Type_1 Interaction

Rule 7 describes that the type_1 expression (i.e. Type_1_Expression) is either a type_1 interaction (i.e. Type_1_Interaction) or a conditional type_1 interaction (i.e. one-or-more-armed conditional expression of Type_1_Interaction).

Rule 7

<Type_1_Expression>   ::= <Type_1_Interaction>
  | <Condition> <Type_1_Interaction>
    {"+" <Condition> <Type_1_Interaction>}

Type_1_Or_2 Expression is either Type_1 or Type_2

Rule 6 describes that the type_1_or_2 expression (i.e. Type_1_Or_2_Expression) is either a type_1 expression (i.e. Type_1_Expression) or a type_2 expression (i.e. Type_2_Expression).

Rule 6
<Type_1_Or_2_Expression> ::=   <Type_1_Expression> | <Type_2_Expression>

Zero or More Expressions either are a Sequential Composition or Consist of a Sequential Composition, Followed by a Type_1_Or_2 Expression, and Followed by Zero or More Expressions

Rule 5 describes that zero or more expressions (i.e. Zero_Or_More_Expressions) either a) are a sequential composition (i.e. ●), or b) consist of a sequential composition (i.e. ●), followed by a type_1_or_2 expression (i.e. Type_1_Or_2_Expression), and followed by zero or more expressions (i.e. Zero_Or_More_Expressions).

Rule 5
<Zero_Or_More_Expressions> ::= "● "<Process_Variable> | "● " <Type_1_Or_2_Expression> <Zero_Or_More_Expressions>

An Interaction Flow Diagram Consists of a Type_1 Expression, Followed by Zero or More Expressions

Rule 4 describes that an interaction flow diagram (i.e. IFD) consists of a type_1 expression (i.e. Type_1_Expression) and followed by zero or more expressions (i.e. Zero_Or_More_Expressions).

Rule 4
<IFD> ::=   <Type_1_Expression> <Zero_Or_More_Expressions>

A NULL Process or the Parallel Composition Composing the Recursion of an Interaction Flow Diagram and the Parallel of All Recursive Interaction Flow Diagrams Defines the Parallel of All Recursive Interaction Flow Diagrams

Rule 3 describes that we use either a) a null process (i.e. *STOP*), or b) the parallel composition (i.e. ‖) composing the recursion (i.e. **fix**) of an interaction flow diagram and the parallel of all recursive interaction flow diagrams (i.e. Parallel_of_FixIFD), to define the parallel of all recursive interaction flow diagrams (i.e. Parallel_of_FixIFD).

Rule 3

<Parallel_of_FixIFD> ::=   *STOP*
  | "**fix**("<Process_Variable>"="<IFD><Process_Variable>")"
    "‖" <Parallel_of_FixIFD>

Parallel of All Recursive Interaction Flow Diagrams Defines the Multi-Queue SBC Process of the Systemic View of Each Level

Rule 2 describes that the parallel of all recursive interaction flow diagrams (i.e. Parallel_of_FixIFD), in which each interaction flow diagram may loop itself a countably infinite times, defines the multi-queue SBC process of the systemic view of each level (i.e. Concept, Analysis, Design, Implementation).

Rule 2
<Multi-Queue_SBC_Process_of_Each_Level's_Systemic_View> ::= <Parallel_of_FixIFD>

Renaming the Multi-Queue SBC Process of Each Level's Systemic View Defines the Structural Composition of Multi-Queue SBC Process of the Systemic View of Each Level

Rule 1 describes that applying the renaming function (i.e. [*f*]) to the multi-queue SBC process of each level's (i.e. Concept, Analysis, Design, Implementation) systemic view defines the structural composition of multi-queue SBC process of the systemic view of each level (i.e. Concept, Analysis, Design, Implementation).

Rule 1

<Structural Composition of Multi-Queue_SBC_Process_of_Each_Level's_Systemic_View> ::=
    <Multi-Queue_SBC_Process_of_Each_Level's_Systemic_View> [*f*]

For each renaming function f, the renaming combinatory $[f]$, postfixed to a process or an interaction, has the effect of renaming the components (of the process or interaction) as dictated by f. We often write $C'_1/C_1,\ldots,\ C'_n/C_n$ for the renaming function for which $f(C_i) = C'_i$ for $i = 1,\ldots,$ n.

After renaming, an interaction may be internalized to be an internal interaction (i.e. λ) if it describes that a component interacts with itself.

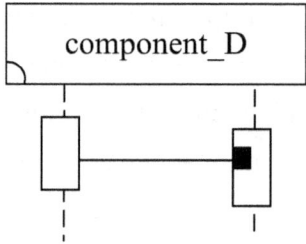

PART V: TRANSITIONAL SEMANTICS OF MULTI-QUEUE SBC PROCESS ALGEBRA

Transitional Semantics

We assume an infinite set Δ of type_1_or_2 interactions, and use a, b to range over Δ. Henceforward we let $\Omega = \Delta \cup \{\lambda\}$, the set of all possible interactions, and use α, β to range over Ω. Further, we let X be the set of process variables, and use X, Y to range over X. We let Φ be the set of process Constants, and use A, B to range over Φ. We let Π be the set of processes, and use P, Q to range over Π. We let Ψ be the set of process expressions, and use E, F to range over Ψ. We let Γ be the set of components, and use C, D to range over Γ.

Entity set	Entity name	Type of entity
Δ	a, b,...	type_1_or_2 interactions
	λ	internal interaction
Ω	$\alpha, \beta,...$	type_1_or_2 or internal interactions
Ω^*	$s,...$	interaction sequences
	$f,...$	renaming functions
X	$X, Y,...$	process variables
Φ	$A, B,...$	process Constants
Π	$P, Q,...$	processes
Ψ	$E, F,...$	process expressions
Γ	$C, D,...$	components
	S	bisimulations

In giving meaning to the multi-queue SBC process algebra, we shall use the following labelled transition system (LTS)

$$(\Psi, \Omega, \rightarrow)$$

which consists of a set Ψ of process expressions, a set Ω of "type_1_or_2 or internal interactions", and a transition relation $\rightarrow \subseteq \Psi \times \Omega \times \Psi$ where $(E_i, \alpha, E_j) \in \rightarrow$ is written as $E_i \xrightarrow{\alpha} E_j$.

The semantics for Ψ consists in the transition rules of each transition relation $\rightarrow$ over $\Psi \times \Omega \times \Psi$. These transition rules will follow the structure of expressions.

We give the complete set of transition rules; the names Prefix, Parallel, Recursion, Rename, and Constant indicate that the rules are associated respectively with Prefix, Parallel Composition, Recursion, and Structural Composition and with Constants.

Prefix	$$\dfrac{}{\alpha \bullet E \xrightarrow{\alpha} E}$$
Parallel$_1$	$$\dfrac{E \xrightarrow{\alpha} E'}{E\|F \xrightarrow{\alpha} E'\|F}$$
Parallel$_2$	$$\dfrac{F \xrightarrow{\alpha} F'}{E\|F \xrightarrow{\alpha} E\|F'}$$
Recursion	$$\dfrac{E\{fix(X=E)/X\} \xrightarrow{\alpha} E'}{fix(X=E) \xrightarrow{\alpha} E'}$$
Rename	$$\dfrac{E \xrightarrow{\alpha} E'}{E[f] \xrightarrow{\alpha[f]} E'[f]}$$
Constant	$$\dfrac{P \xrightarrow{\alpha} P'}{A \xrightarrow{\alpha} P'} \quad (A \overset{def}{=} P)$$

Rule of Prefix

The rule for Prefix can be read as follows: Under any circumstances, we always infer $\alpha \bullet E \xrightarrow{\alpha} E$. That is, an expression, with an interaction prefixed to it, will use this interaction to accomplish the transition.

$$\frac{}{\alpha \bullet E \xrightarrow{\alpha} E}$$

Rules of Parallel Composition

There are two transition rules for parallel composition. Rule Parallel$_1$ indicates that from $E \xrightarrow{\alpha} E'$ we shall infer $E\|F \xrightarrow{\alpha} E'\|F$.

$$\frac{E \xrightarrow{\alpha} E'}{E\|F \xrightarrow{\alpha} E'\|F}$$

Rule Parallel$_2$ indicates that from $F \xrightarrow{\alpha} F'$ we shall infer $E\|F \xrightarrow{\alpha} E\|F'$.

$$\frac{F \xrightarrow{\alpha} F'}{E\|F \xrightarrow{\alpha} E\|F'}$$

Rule of Recursion

The rule for Recursion can be read as follows: This says that any interaction which may be inferred for the **fix** expression 'unwound' once (by substituting itself for its bound variable) may be inferred for the **fix** expression itself.

$$\frac{E\{\mathbf{fix}(X{=}E)/X\} \overset{\alpha}{\to} E'}{\mathbf{fix}(X{=}E) \overset{\alpha}{\to} E'}$$

Rule of Structural Composition

The rule for Structural Composition can be read as follows:

Rule Rename indicates that from $E \xrightarrow{\alpha} E'$ we shall infer

$E[f] \xrightarrow{\alpha[f]} E'[f]$.

$$\frac{E \xrightarrow{\alpha} E'}{E[f] \xrightarrow{\alpha[f]} E'[f]}$$

Rule of Constants

The rule for Constants can be read as follows: the rule of Constants asserts that each Constant has the same transitions as its defining expression.

$$\frac{P \xrightarrow{\alpha} P'}{A \xrightarrow{\alpha} P'} \quad (A \overset{\text{def}}{=} P)$$

Examples of Transitional Semantics

As a first example, consider the multi-queue SBC process Constant A_{01} is defined as "$\mathbf{fix}(X_1{=}a_1{\bullet}X_1)\|\mathbf{fix}(X_2{=}a_2{\bullet}X_2)\|\mathbf{fix}(X_3{=}a_3{\bullet}\lambda{\bullet}X_3)$". The following transition graph shows the semantics of process A_{01}.

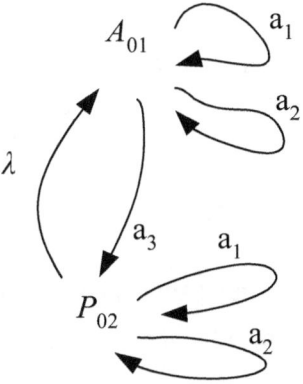

In the transition graph of the A_{01}'s multi-queue SBC process, processes A_{01} and P_{02} are defined as:

$$A_{01} \stackrel{\text{def}}{=\joinrel=} a_1 \bullet A_{01} \ + \ a_2 \bullet A_{01} \ + \ a_3 \bullet P_{02}$$

$$P_{02} \stackrel{\text{def}}{=\joinrel=} a_1 \bullet P_{02} \ + \ a_2 \bullet P_{02} \ + \ \lambda \bullet A_{01}$$

As a second example, consider the multi-queue SBC process Constant B_{01} is defined as "$\mathbf{fix}(X_4=b_1\bullet b_2\bullet X_4)\|fix(X_5=\lambda\bullet b_3\bullet X_5)$". The following transition graph shows the semantics of process B_{01}.

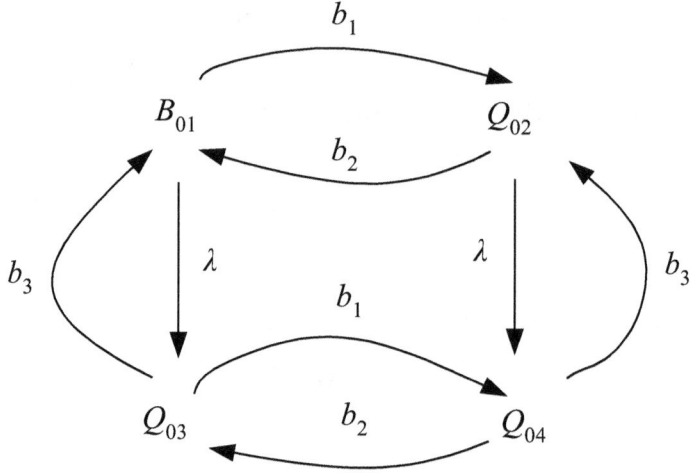

In the transition graph of the B_{01}'s multi-queue SBC process, processes B_{01}, Q_{02}, Q_{03}, and Q_{04} are defined as:

$$B_{01} \overset{\text{def}}{=\!=} b_1\bullet Q_{02} + \lambda\bullet Q_{03}$$

$$Q_{02} \overset{\text{def}}{=\!=} b_2\bullet B_{01} + \lambda\bullet Q_{04}$$

$$Q_{03} \overset{\text{def}}{=\!=} b_1\bullet Q_{04} + b_3\bullet B_{01}$$

$$Q_{04} \overset{\text{def}}{=\!=} b_2\bullet Q_{03} + b_3\bullet Q_{02}$$

As a third example, consider the multi-queue SBC process Constant C_{01} is defined as "$\mathbf{fix}(X_6 = c_1 \bullet c_2 \bullet X_6) \| \mathbf{fix}(X_7 = c_3 \bullet c_4 \bullet \lambda \bullet X_7)$". The following transition graph shows the semantics of process C_{01}.

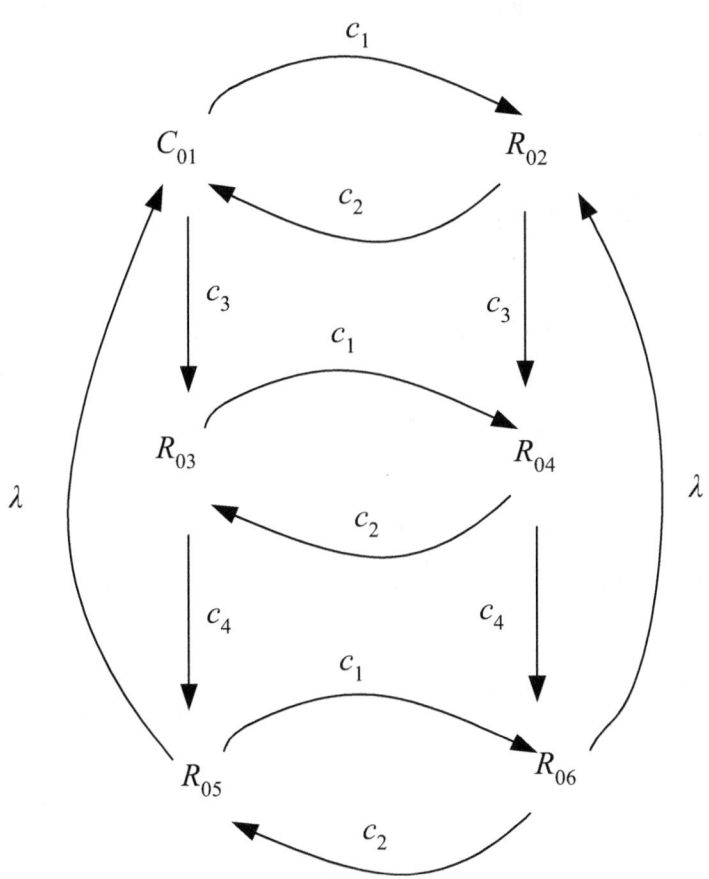

In the transition graph of the C_{01}'s multi-queue SBC process, processes C_{01}, R_{02}, R_{03}, R_{04}, R_{05}, and R_{06} are defined as:

$$C_{01} \stackrel{\text{def}}{=} c_1 \bullet R_{02} + c_3 \bullet R_{03}$$

$$R_{02} \stackrel{\text{def}}{=} c_2 \bullet C_{01} + c_3 \bullet R_{04}$$

$$R_{03} \stackrel{\text{def}}{=} c_1 \bullet R_{04} + c_4 \bullet R_{05}$$

$$R_{04} \stackrel{\text{def}}{=} c_2 \bullet R_{03} + c_4 \bullet R_{06}$$

$$R_{05} \stackrel{\text{def}}{=} c_1 \bullet R_{06} + \lambda \bullet C_{01}$$

$$R_{06} \stackrel{\text{def}}{=} c_2 \bullet R_{05} + \lambda \bullet R_{02}$$

PART VI: OBSERVATION CONGRUENCE THEORY OF MULTI-QUEUE SBC PROCESS ALGEBRA

Preliminary Definitions

A few preliminary definitions are needed. Definition 6.1 and Definition 6.2 are easy to comprehend.

Definition 6.1 $\qquad \widehat{\alpha} = \varepsilon$ (the empty sequence), if $\alpha = \lambda$

$\qquad\qquad\qquad = \alpha$, if $\alpha \neq \lambda$

Definition 6.2 $\quad E \overset{\alpha}{\Longrightarrow} E'$ iff $E (\overset{\lambda}{\rightarrow})^* \overset{\alpha}{\rightarrow} (\overset{\lambda}{\rightarrow})^* E'$

$\qquad\qquad\quad E \overset{\varepsilon}{\Longrightarrow} E'$ iff $E (\overset{\lambda}{\rightarrow})^* (\overset{\lambda}{\rightarrow})^* E'$

Observation Equivalence

To achieve the observation congruence, we need to define the observation equivalence first. Definition 6.3 and Definition 6.4 together define the observation equivalence.

Definition 6.3 A binary relation $S \subseteq \Pi \times \Pi$ over multi-queue SBC processes is a *bisimulation* if $(P, Q) \in S$ implies, for all or each $\alpha \in \Omega$,

(i) Whenever $P \xrightarrow{\alpha} P'$ then, for some Q', $Q \xRightarrow{\hat{\alpha}} Q'$ and $(P', Q') \in S$

(ii) Whenever $Q \xrightarrow{\alpha} Q'$ then, for some P', $P \xRightarrow{\hat{\alpha}} P'$ and $(P', Q') \in S$

Definition 6.4 P and Q are observation equivalent, written $P \approx Q$, if $(P, Q) \in S$ for some bisimulation S. That is,

$$\approx = \bigcup (S : S \text{ is a bisimulation})$$

Observation Congruence

Once we have the definition of observation equivalence, i.e. $\approx$, we shall use it to define the observation congruence.

Definition 6.5 P and Q are observation congruent, written $P = Q$, if for all α

(i) Whenever $P \xrightarrow{\alpha} P'$ then, for some Q', $Q \xRightarrow{\alpha} Q'$ and $P' \approx Q'$;

(ii) Whenever $Q \xrightarrow{\alpha} Q'$ then, for some P', $P \xRightarrow{\alpha} P'$ and $P' \approx Q'$.

Examples of Observation Congruence Verification

As a first example, consider the multi-queue SBC process P_{01} is syntactically represented as "$\mathbf{fix}(X_{01}=a_1 \bullet X_{01}) \| \mathbf{fix}(X_{02}=a_2 \bullet X_{02}) \| \mathbf{fix}(X_{03}=a_3 \bullet X_{03})$" and the multi-queue SBC process Q_{01} is syntactically represented as "$\mathbf{fix}(X_{04}=a_1 \bullet X_{04}) \| \mathbf{fix}(X_{05}=a_2 \bullet X_{05}) \| \mathbf{fix}(X_{06}=a_3 \bullet \lambda \bullet X_{06})$", which are represented by the following transition graphs.

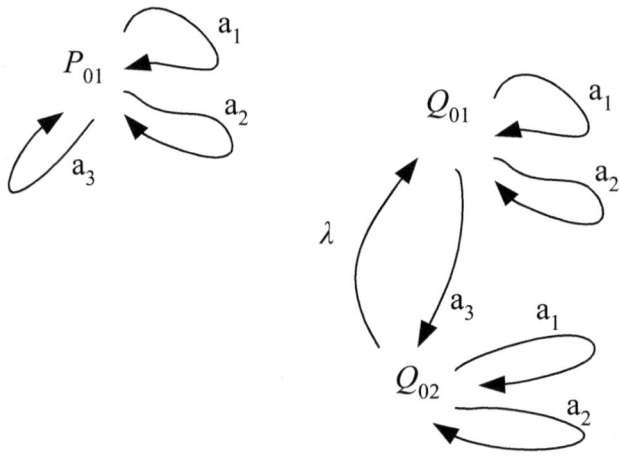

We can easily verify that $S = \{(P_{01}, Q_{01}), (P_{01}, Q_{02})\}$ is a bisimulation.

Using the S bisimulation, we then are able to verify that P_{01} and Q_{01} are observation congruent because (1) $P_{01} \xrightarrow{a_1} P_{01}$, then we have Q_{01} that $Q_{01} \xRightarrow{a_1} Q_{01}$ and $P_{01} \approx Q_{01}$, and (2) $P_{01} \xrightarrow{a_2} P_{01}$, then we have Q_{01} that $Q_{01} \xRightarrow{a_2} Q_{01}$ and $P_{01} \approx Q_{01}$, and (3) $P_{01} \xrightarrow{a_3} P_{01}$, then we have Q_{02} that $Q_{01} \xRightarrow{a_3} Q_{02}$ and $P_{01} \approx Q_{02}$, and (4) $Q_{01} \xrightarrow{a_1} Q_{01}$, then we have P_{01} that $P_{01} \xRightarrow{a_1} P_{01}$ and $P_{01} \approx Q_{01}$, and (5) $Q_{01} \xrightarrow{a_2} Q_{01}$, then we have P_{01} that $P_{01} \xRightarrow{a_2} P_{01}$ and $P_{01} \approx Q_{01}$, and (6) $Q_{01} \xrightarrow{a_3} Q_{02}$, then we have P_{01} that $P_{01} \xRightarrow{a_3} P_{01}$ and $P_{01} \approx Q_{02}$.

As a second example, consider the multi-queue SBC process P_{03} is syntactically represented as "**fix**$(X_{07}=a_4 \bullet X_{07})\|$**fix**$(X_{08}=a_5 \bullet X_{08})$" and the multi-queue SBC process Q_{03} is syntactically represented as "**fix**$(X_{09}=a_4 \bullet X_{09})\|$**fix**$(X_{10}=\lambda \bullet a_5 \bullet X_{10})$", which are represented by the following transition graphs.

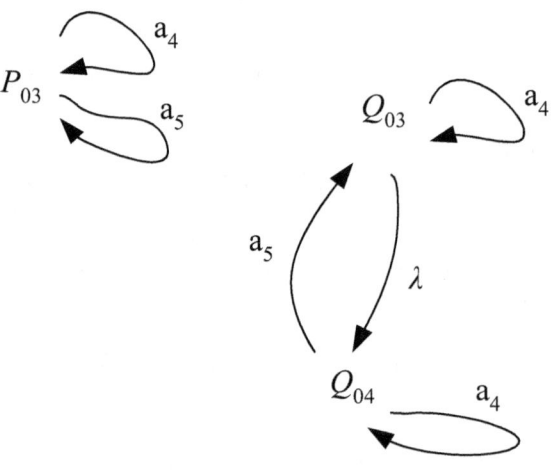

We can easily verify that $S = \{(P_{03}, Q_{03}), (P_{03}, Q_{04})\}$ is a bisimulation.

Using the S bisimulation, we then are able to verify that P_{03} and Q_{03} are not observation congruent because for $Q_{03} \xrightarrow{\lambda} Q_{04}$ and we find no P_{0m} that $P_{03} \overset{\lambda}{\Longrightarrow} P_{0m}$ and $P_{0m} \approx Q_{04}$.

As a third example, consider the multi-queue SBC process P_{05} is syntactically represented as "$\mathbf{fix}(X_{11}=a_6 \bullet X_{11}) \| \mathbf{fix}(X_{12}=a_7 \bullet X_{12})$" and the multi-queue SBC process Q_{05} is syntactically represented as "$\mathbf{fix}(X_{13}=a_6 \bullet X_{13}) \| \mathbf{fix}(X_{14}=a_7 \bullet \lambda \bullet X_{14})$", which are represented by the following transition graphs.

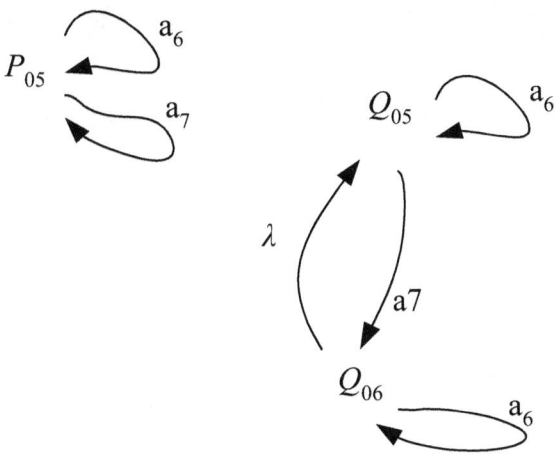

We can easily verify that $S = \{(P_{05}, Q_{05}), (P_{05}, Q_{06})\}$ is a bisimulation.

Using the S bisimulation, we then are able to verify that P_{05} and Q_{06} are observation congruent because (1) $P_{05} \xrightarrow{a_6} P_{05}$, then we have Q_{05} that $Q_{05} \xRightarrow{a_6} Q_{05}$ and $P_{05} \wasytilde Q_{05}$, and (2) $P_{05} \xrightarrow{a_7} P_{05}$, then we have Q_{06} that $Q_{05} \xRightarrow{a_7} Q_{06}$ and $P_{05} \wasytilde Q_{06}$, and (3) $Q_{05} \xrightarrow{a_6} Q_{01}$, then we have P_{05} that $P_{05} \xRightarrow{a_6} P_{05}$ and $P_{05} \wasytilde Q_{05}$, and (4) $Q_{05} \xrightarrow{a_7} Q_{06}$, then we have P_{05} that $P_{05} \xRightarrow{a_7} P_{05}$ and $P_{05} \wasytilde Q_{06}$.

PART VII: MULTI-LEVEL VIEW OF SYSTEMS ARCHITECTURE

SBC View Model of Systems Architecture

In the SBC view model of systems architecture, dimension 2 stands for the multi-level view which contains the concept, analysis, design, and implementation views; dimension 3 stands for the systemic view which contains the collection of all interaction flow diagrams.

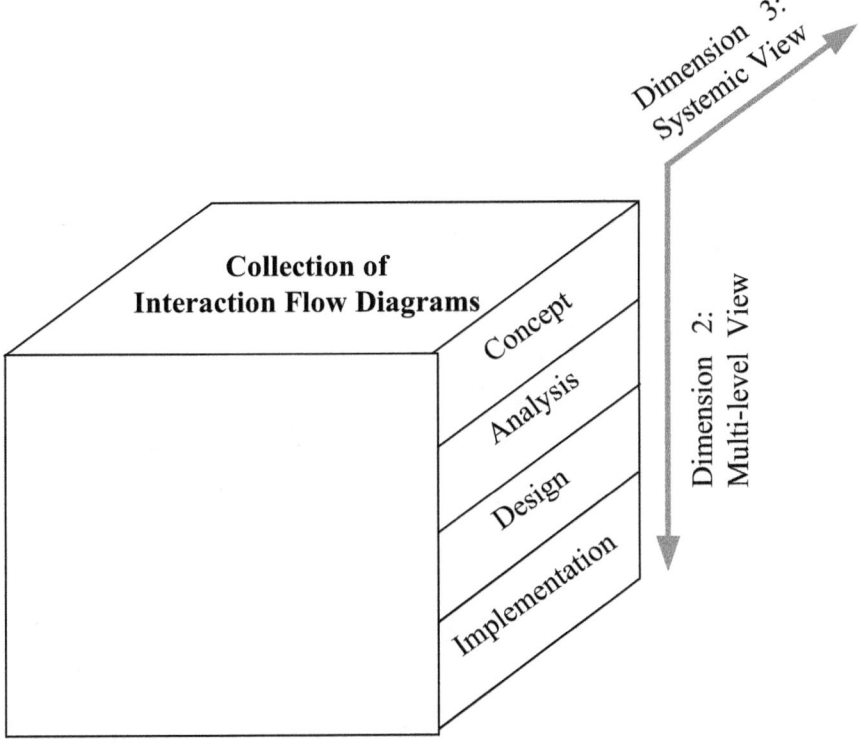

In the SBC multi-level view of systems architecture, analysis view is one level down structural decomposition (with observation congruence verification) of the concept view; design view is one level down structural decomposition (with observation congruence verification) of the analysis view; implementation view is one level down structural decomposition (with observation congruence verification) of the design view.

Analysis View is One Level down Structural Decomposition of the Concept View

To demonstrate that the analysis view is one level down structural decomposition (with observation congruence verification) of the concept view, we need to go through the following four steps:

(A) Get the multi-queue SBC process of the concept view.

(B) Get the multi-queue SBC process of the analysis view.

(C) Get the multi-queue SBC process of the structural composition of the analysis view.

(D) Verify that there is observation congruence of "the concept view" and "the structural composition of the analysis view".

Design View is One Level down Structural Decomposition of the Analysis View

To demonstrate that the design view is one level down structural decomposition (with observation congruence verification) of the analysis view, we need to go through the following four steps:

(A) Get the multi-queue SBC process of the analysis view.

(B) Get the multi-queue SBC process of the design view.

(C) Get the multi-queue SBC process of the structural composition of the design view.

(D) Verify that there is observation congruence of "the analysis view" and "the structural composition of the design view".

Implementation View is One Level down Structural Decomposition of the Design View

To demonstrate that the implementation view is one level down structural decomposition (with observation congruence verification) of the design view, we need to go through the following four steps:

(A) Get the multi-queue SBC process of the design view.

(B) Get the multi-queue SBC process of the implementation view.

(C) Get the multi-queue SBC process of the structural composition of the implementation view.

(D) Verify that there is observation congruence of "the design view" and "the structural composition of the implementation view".

PART VIII: CASE STUDY --
KURDI UNIVERSITY

SBC View Model of Kurdi University

In the SBC view model of *Kurdi University*, dimension 2 stands for the multi-level view which contains the concept, analysis, design, and implementation views; dimension 3 stands for the systemic view which contains the collection of all interaction flow diagrams.

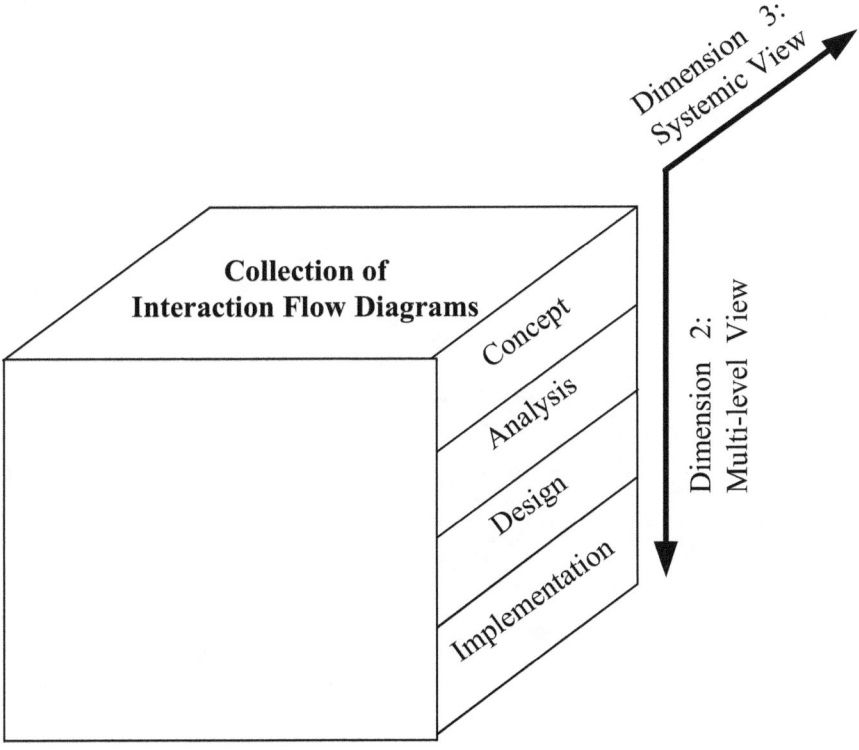

In the SBC multi-level view of *Kurdi University*, analysis view is one level down structural decomposition (with observation congruence verification) of the concept view; design view is one level down structural decomposition (with observation congruence verification) of the analysis view; implementation view is one level down structural decomposition (with observation congruence verification) of the design view.

Multi-Queue SBC Process of the Concept View

We draw the Architecture Hierarchy Diagram (AHD) of the concept view of *Kurdi University* as follows:

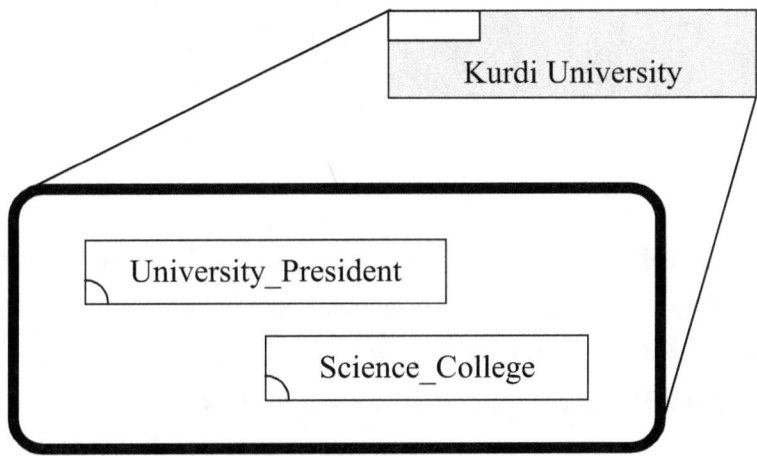

The overall behavior of the concept view of *Kurdi University* includes two behaviors: *Study_Calculus_Course* and *Study_Algebra_Course*. Each of them is described by an individual IFD.

An IFD of the *Study_Calculus_Course* behavior is shown below. First, actor *Student* interacts with the *University_President* component through the *University_Teach_Calculus* operation call interaction. Next, component *University_President* interacts with the *Science_College* component through the

College_Teach_Calculus operation call interaction.

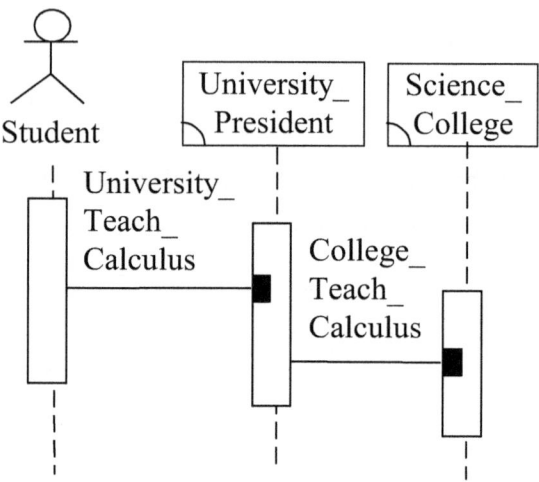

An IFD of the *Study_Algebra_Course* behavior is shown below. First, actor *Student* interacts with the *University_President* component through the *University_Teach_Algebra* operation call interaction. Next, component *University_President* interacts with the *Science_College* component through the *College_Teach_Algebra* operation call interaction

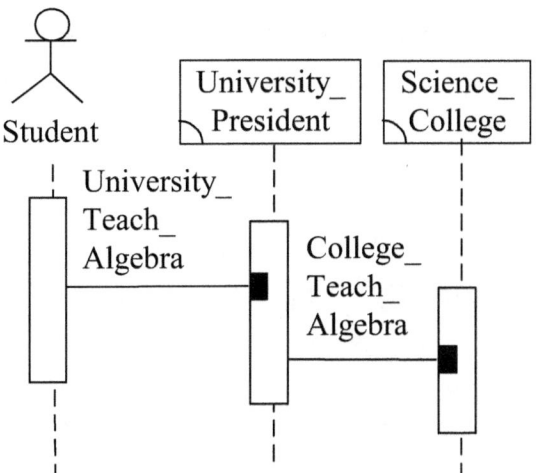

We draw the multi-queue SBC process algebra Backus-Naur Form tree of the concept view of *Kurdi University* as follows:

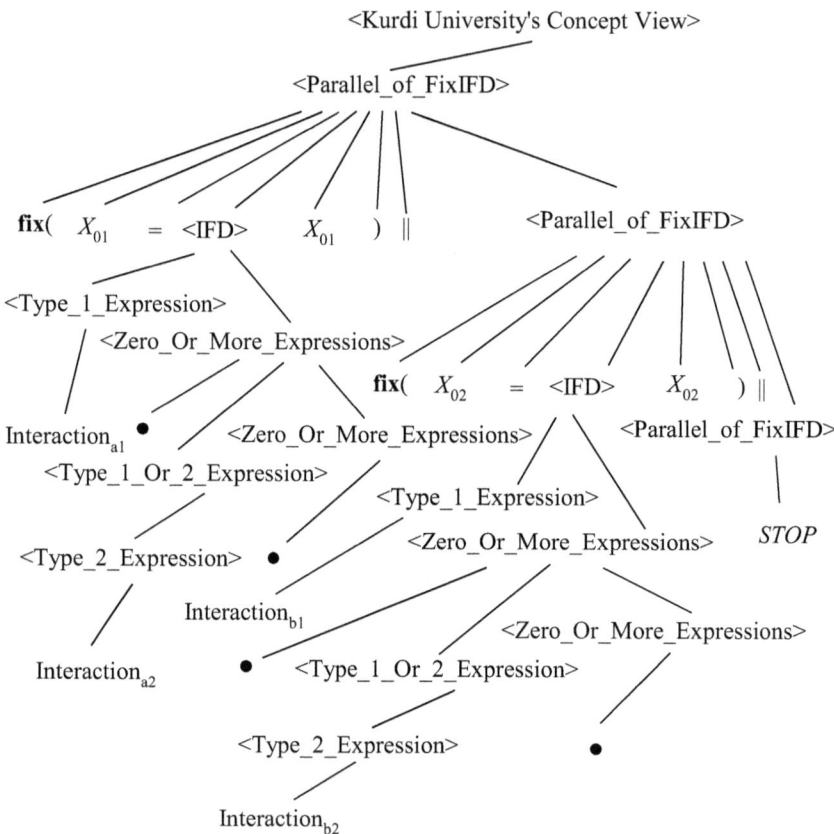

Interaction$_{a1}$ stands for the 1st interaction of the ath interaction flow diagram of the concept view of *Kurdi University*. Interaction$_{a1}$ is a type_1 interaction which describes the *Student* actor interacts with the *University_President* component.

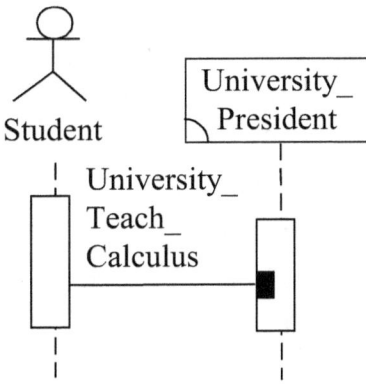

Interaction$_{a2}$ stands for the 2nd interaction of the ath interaction flow diagram of the concept view of *Kurdi University*. Interaction$_{a2}$ is a type_2 interaction which describes the *University_President* component interacts with the *Science_College* component.

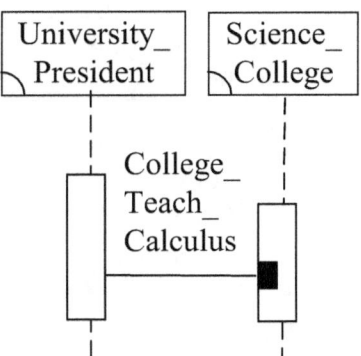

Interaction$_{b1}$ stands for the 1st interaction of the bth interaction flow diagram of the concept view of *Kurdi University*. Interaction$_{b1}$ is a type_1 interaction which describes the *Student* actor interacts with the *University_President* component.

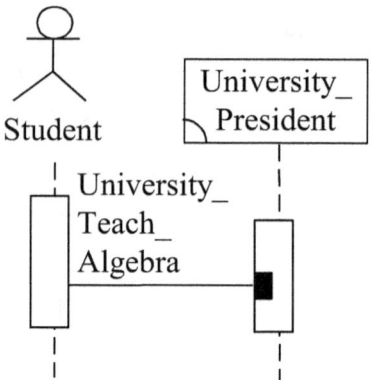

Interaction$_{b2}$ stands for the 2nd interaction of the bth interaction flow diagram of the concept view of *Kurdi University*. Interaction$_{b2}$ is a type_2 interaction which describes the *University_President* component interacts with the *Science_College* component.

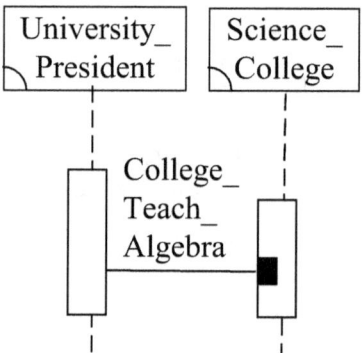

FixIFD$_a$ describes the recursion of the ath interaction flow diagram, i.e. *Study_Calculus_Course* behavior, of the concept view of *Kurdi University*. FixIFD$_a$ is syntactically represented as "**fix**(X_{01}=Interaction$_{a1}$•Interaction$_{a2}$•X_{01})".

$$\text{FixIFD}_a \;\overset{\text{def}}{=\!=}$$

$$\mathbf{fix}(X_{01}=\text{Interaction}_{a1}\bullet\;\text{Interaction}_{a2}\bullet X_{01})$$

FixIFD_b describes the recursion of the bth interaction flow diagram, i.e. *Study_Algebra_Course* behavior, of the concept view of *Kurdi University*. FixIFD_b is syntactically represented as "$\mathbf{fix}(X_{02}=\text{Interaction}_{b1}\bullet\text{Interaction}_{b2}\bullet X_{02})$".

$$\text{FixIFD}_b \;\overset{\text{def}}{=\!=}$$

$$\mathbf{fix}(X_{02}=\text{Interaction}_{b1}\bullet\;\text{Interaction}_{b2}\bullet X_{02})$$

Multi-queue SBC process of the concept view of *Kurdi University* is syntactically represented as $\text{FixIFD}_a\|\text{FixIFD}_b$ which equals to "$\mathbf{fix}(X_{01}=\text{Interaction}_{a1}\bullet\text{Interaction}_{a2}\bullet X_{01})\;\|\;\mathbf{fix}(X_{02}=\text{Interaction}_{b1}\bullet\text{Interaction}_{b2}\bullet X_{02})$".

$$\text{Kurdi University's Concept View}\;\overset{\text{def}}{=\!=}$$

$$\mathbf{fix}(X_{01}=\text{Interaction}_{a1}\bullet\;\text{Interaction}_{a2}\bullet X_{01})\;\|$$
$$\mathbf{fix}(X_{02}=\text{Interaction}_{b1}\bullet\;\text{Interaction}_{b2}\bullet X_{02})$$

Multi-Queue SBC Process of the Analysis View

We draw the Architecture Hierarchy Diagram (AHD) of the analysis view of *Kurdi University* as follows:

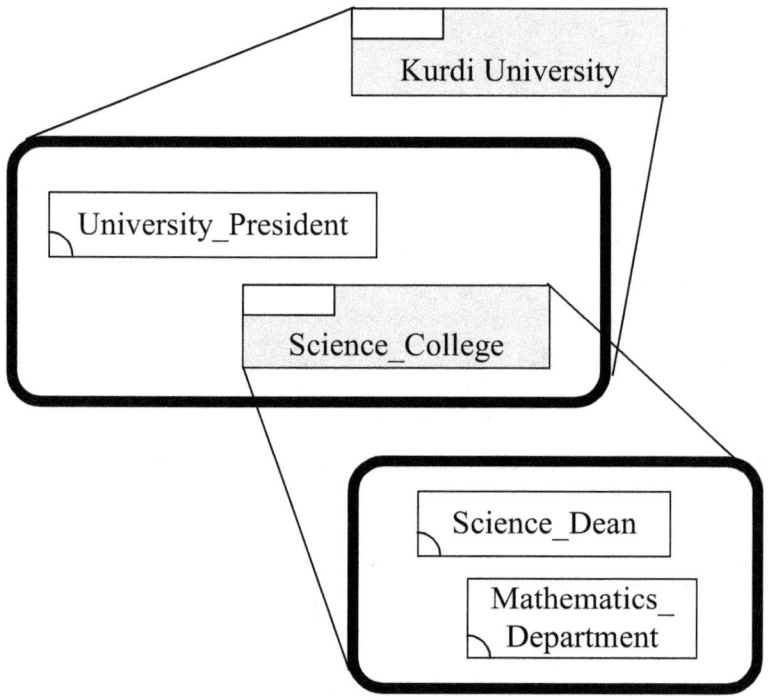

The overall behavior of the analysis view of *Kurdi University* includes two behaviors: *Study_Calculus_Course* and *Study_Algebra_Course*. Each of them is described by an individual IFD.

An IFD of the *Study_Calculus_Course* behavior is shown below. First, actor *Student* interacts with the *University_President* component through the *University_Teach_Calculus* operation call interaction. Next, component *University_President* interacts with the *Science_Dean* component through the *College_Teach_Calculus* operation call interaction. Finally, component *Science_Dean* interacts with the *Mathematics_Department* component through the *Department_Teach_Calculus* operation call interaction.

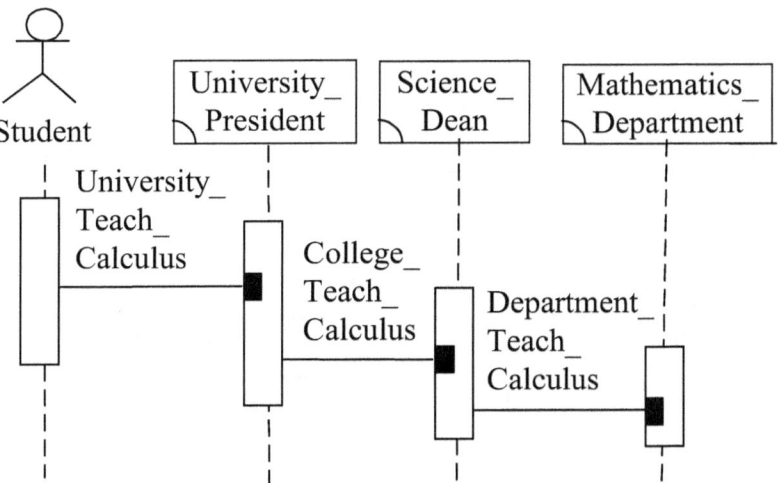

An IFD of the *Study_Algebra_Course* behavior is shown below. First, actor *Student* interacts with the *University_President* component through the *University_Teach_Algebra* operation call interaction. Next, component *University_President* interacts with the *Science_Dean* component through the *College_Teach_Algebra* operation call interaction. Finally, component *Science_Dean* interacts with the *Mathematics_Department* component through the *Department_Teach_Algebra* operation call interaction.

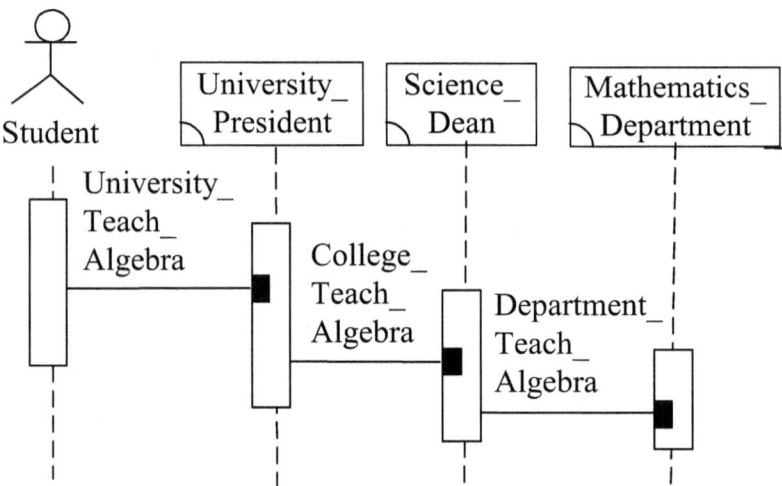

We draw the multi-queue SBC process algebra Backus-Naur Form tree of the analysis view of *Kurdi University* as follows:

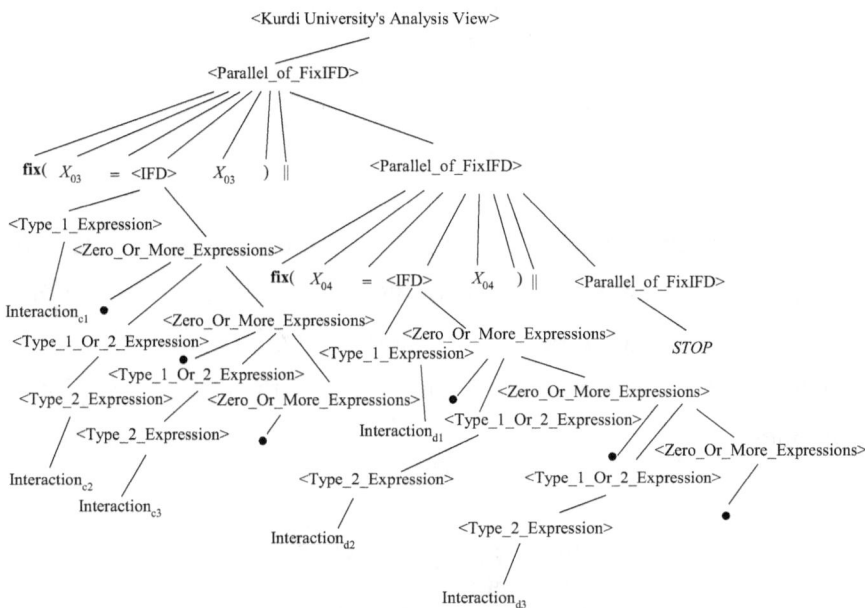

Interaction$_{c1}$ stands for the 1st interaction of the cth interaction flow diagram of the analysis view of *Kurdi University*. Interaction$_{c1}$ is a type_1 interaction which describes the *Student* actor interacts with the *University_President* component.

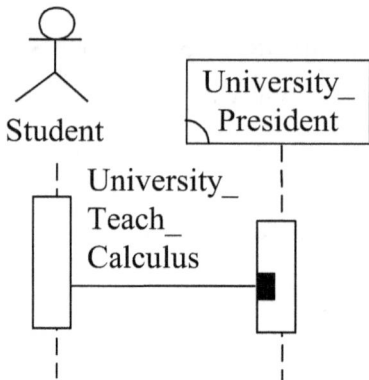

Interaction$_{c2}$ stands for the 2nd interaction of the cth interaction flow diagram of the analysis view of *Kurdi University*. Interaction$_{c2}$ is a type_2 interaction which describes the *University_President* component interacts with the *Science_Dean* component.

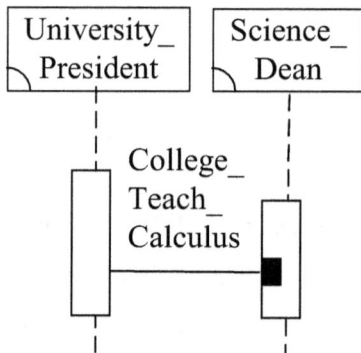

Interaction$_{c3}$ stands for the 3rd interaction of the cth interaction flow diagram of the analysis view of *Kurdi University*. Interaction$_{c3}$ is a type_2 interaction which describes the *Science_Dean* component interacts with the *Mathematics_Department* component.

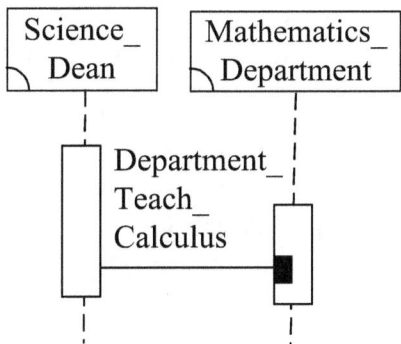

Interaction$_{d1}$ stands for the 1st interaction of the dth interaction flow diagram of the analysis view of *Kurdi University*. Interaction$_{d1}$ is a type_1 interaction which describes the *Student* actor interacts with the *University_President* component.

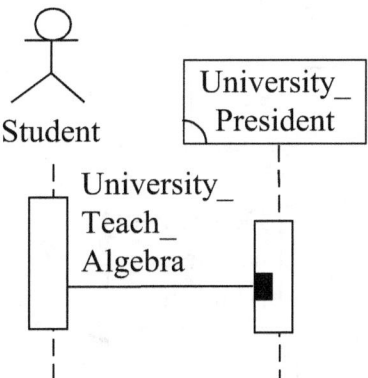

Interaction$_{d2}$ stands for the 2nd interaction of the dth interaction flow diagram of the analysis view of *Kurdi University*. Interaction$_{d2}$ is a type_2 interaction which describes the *University_President* component interacts with the *Science_Dean* component.

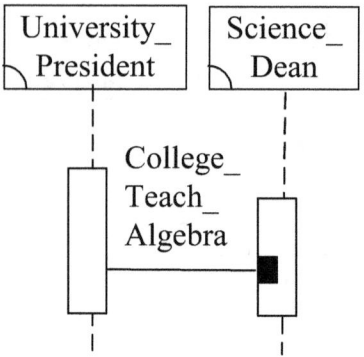

Interaction$_{d3}$ stands for the 3rd interaction of the dth interaction flow diagram of the analysis view of *Kurdi University*. Interaction$_{d3}$ is a type_2 interaction which describes the *Science_Dean* component interacts with the *Mathematics_Department* component.

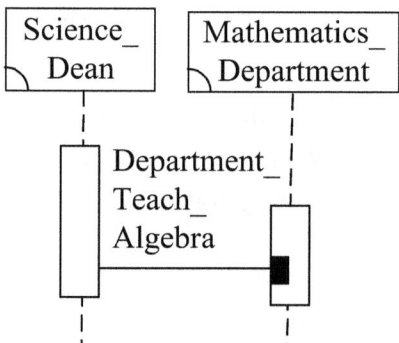

FixIFD$_c$ describes the recursion of the cth interaction flow diagram, i.e. *Study_Calculus_Course* behavior, of the analysis view of *Kurdi University*. FixIFD$_c$ is syntactically represented as "**fix**(X_{03}=Interaction$_{c1}$●Interaction$_{c2}$●Interaction$_{c3}$●X_{03})".

$$\text{FixIFD}_c \stackrel{\text{def}}{=\!=}$$

$$\mathbf{fix}(X_{03} = \text{Interaction}_{c1} \bullet \text{Interaction}_{c2} \bullet \text{Interaction}_{c3} \bullet X_{03})$$

FixIFD$_d$ describes the recursion of the dth interaction flow diagram, i.e. *Study_Algebra_Course* behavior, of the analysis view of *Kurdi University*. FixIFD$_d$ is syntactically represented as "$\mathbf{fix}(X_{04}=\text{Interaction}_{d1} \bullet \text{Interaction}_{d2} \bullet \text{Interaction}_{d3} \bullet X_{04})$".

$$\text{FixIFD}_d \stackrel{\text{def}}{=\!=}$$

$$\mathbf{fix}(X_{04} = \text{Interaction}_{d1} \bullet \text{Interaction}_{d2} \bullet \text{Interaction}_{d3} \bullet X_{04})$$

Multi-queue SBC process of the analysis view of *Kurdi University* is syntactically represented as FixIFD$_c$||FixIFD$_d$ which equals to "$\mathbf{fix}(X_{03}=\text{Interaction}_{c1} \bullet \text{Interaction}_{c2} \bullet \text{Interaction}_{c3} \bullet X_{03})$ || $\mathbf{fix}(X_{04}=\text{Interaction}_{d1} \bullet \text{Interaction}_{d2} \bullet \text{Interaction}_{d3} \bullet X_{04})$".

$$\text{Kurdi University's Analysis View} \stackrel{\text{def}}{=\!=}$$

$$\mathbf{fix}(X_{03}=\text{Interaction}_{c1} \bullet \text{Interaction}_{c2} \bullet \text{Interaction}_{c3} \bullet X_{03}) \ ||$$
$$\mathbf{fix}(X_{04}=\text{Interaction}_{d1} \bullet \text{Interaction}_{d2} \bullet \text{Interaction}_{d3} \bullet X_{04})$$

Multi-Queue SBC Process of the Structural Composition of the Analysis View

Structural composition of the analysis view of *Kurdi University* means to compose the *Science_Dean* and *Mathematics_Department* components into the *Science_College* component. That is, we will rename the *Science_Dean* component to the *Science_College* component; we also will rename the *Mathematics_Department* component to the *Science_College* component.

[Science_College/Science_Dean,
Science_College/Mathematics_Department]

We draw the multi-queue SBC process algebra Backus-Naur Form tree of the structural composition of the analysis view of *Kurdi University* as follows:

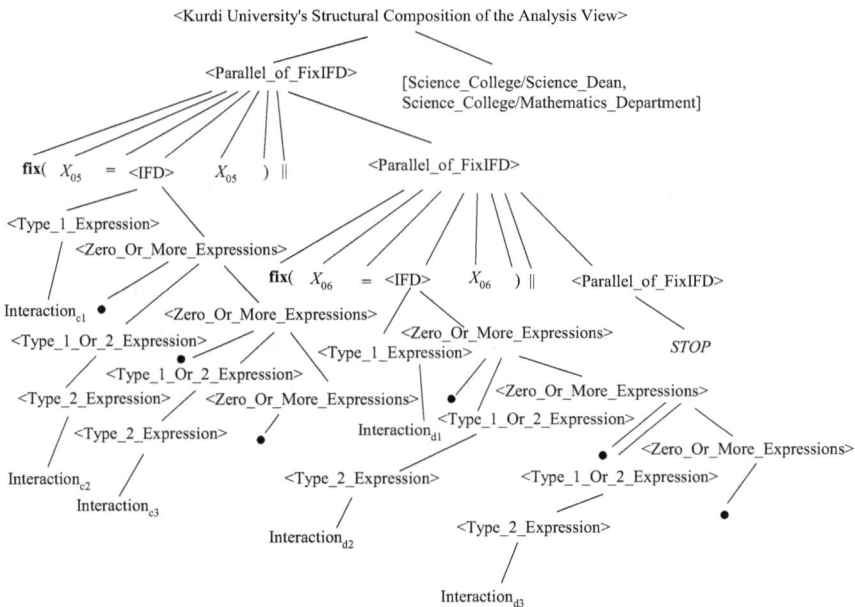

Interaction$_{e1}$=Interaction$_{c1}$[Science_College/Science_Dean,Science_College/Mathematics_Department]=Interaction$_{a1}$ stands for the 1st interaction of the eth interaction flow diagram of the structural composition of the analysis view of *Kurdi University*. Interaction$_{e1}$ is a type_1 interaction which describes the *Student* actor interacts with the *University_President* component.

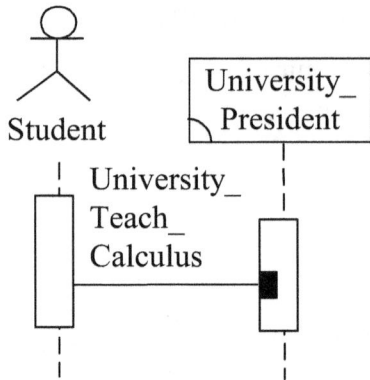

Interaction$_{e2}$=Interaction$_{c2}$[Science_College/Science_Dean,Science_College/Mathematics_Department]=Interaction$_{a2}$ stands for the 2nd interaction of the eth interaction flow diagram of the structural composition of the analysis view of *Kurdi University*. Interaction$_{e2}$ is a type_2 interaction which describes the *University_President* component interacts with the *Science_College* component.

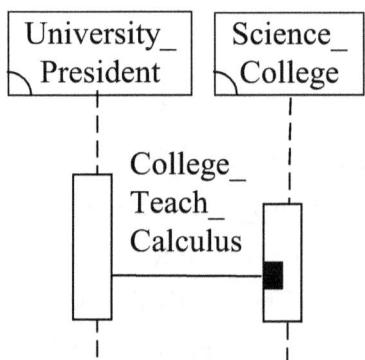

Interaction$_{e3}$=Interaction$_{c3}$[Science_College/Science_Dean,Science_College/Mathematics_Department] stands for the 3rd interaction of the eth interaction flow diagram of the structural composition of the analysis view of *Kurdi University*. Interaction$_{e3}$ is an internal interaction (i.e. λ) inside the *Science_College* component.

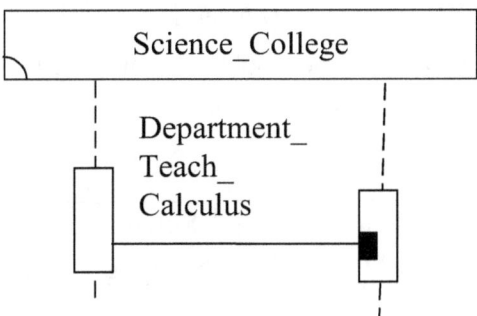

Interaction$_{f1}$=Interaction$_{d1}$[Science_College/Science_Dean,Science_College/Mathematics_Department]=Interaction$_{b1}$ stands for the 1st interaction of the fth interaction flow diagram of the structural composition of the analysis view of *Kurdi University*. Interaction$_{f1}$ is a type_1 interaction which describes the *Student* actor interacts with the *University_President* component.

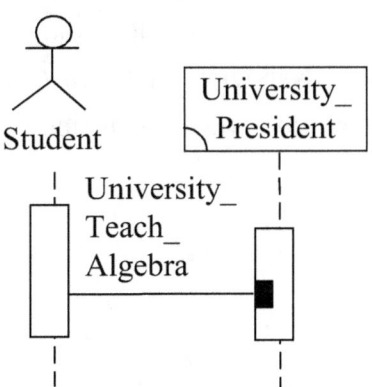

Interaction$_{f2}$=Interaction$_{d2}$[Science_College/Science_Dean,S cience_College/Mathematics_Department]=Interaction$_{b2}$ stands for the 2nd interaction of the fth interaction flow diagram of the structural composition of the analysis view of *Kurdi University*. Interaction$_{f2}$ is a type_2 interaction which describes the *University_President* component interacts with the *Science_College* component.

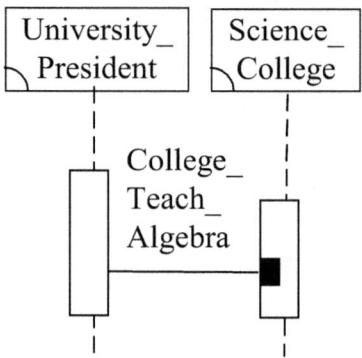

Interaction$_{f3}$=Interaction$_{d3}$[Science_College/Science_Dean,S cience_College/Mathematics_Department] stands for the 3rd interaction of the fth interaction flow diagram of the structural composition of the analysis view of *Kurdi University*. Interaction$_{f3}$ is an internal interaction (i.e. λ) inside the *Science_College* component.

FixIFD$_e$ describes the recursion of the eth interaction flow diagram, i.e. *Study_Calculus_Course* behavior, of the structural composition of the analysis view of *Kurdi University*. FixIFD$_e$ is syntactically represented as "**fix**$(X_{05}$=Interaction$_{a1}$•Interaction$_{a2}$•λ•$X_{05})$".

$$\text{FixIFD}_e \stackrel{\text{def}}{=\!=}$$

$$\textbf{fix}(X_{05}\text{=Interaction}_{a1} \bullet \text{Interaction}_{a2} \bullet \lambda \bullet X_{05})$$

FixIFD$_f$ describes the recursion of the fth interaction flow diagram, i.e. *Study_Algebra_Course* behavior, of the structural composition of the analysis view of *Kurdi University*. FixIFD$_f$ is syntactically represented as "**fix**$(X_{06}$=Interaction$_{b1}$•Interaction$_{b2}$•λ•$X_{06})$".

$$\text{FixIFD}_f \stackrel{\text{def}}{=\!=}$$

$$\textbf{fix}(X_{06}\text{=Interaction}_{b1} \bullet \text{Interaction}_{b2} \bullet \lambda \bullet X_{06})$$

Multi-queue SBC process of the structural composition of the analysis view of *Kurdi University* is syntactically represented as FixIFD$_e$||FixIFD$_f$ which equals to "**fix**(X_{05}=Interaction$_{a1}$●Interaction$_{a2}$●λ●X_{05}) ||
fix(X_{06}=Interaction$_{b1}$●Interaction$_{b2}$●λ●X_{06})".

Kurdi University's Structural Composition of the Analysis View $\overset{def}{=\!=}$

fix(X_{05}=Interaction$_{a1}$ ● Interaction$_{a2}$● λ ●X_{05}) ||
fix(X_{06}=Interaction$_{b1}$ ● Interaction$_{b2}$● λ ●X_{06})

Observation Congruence of "the Concept View" and "the Structural Composition of the Analysis View"

We syntactically represent multi-queue SBC process P_{01} as $FixIFD_a \| FixIFD_b$ which equals to "**fix**$(X_{01}=Interaction_{a1} \bullet Interaction_{a2} \bullet X_{01})$ $\|$ **fix**$(X_{02}=Interaction_{b1} \bullet Interaction_{b2} \bullet X_{02})$". The following transition graph shows the semantics of process P_{01}.

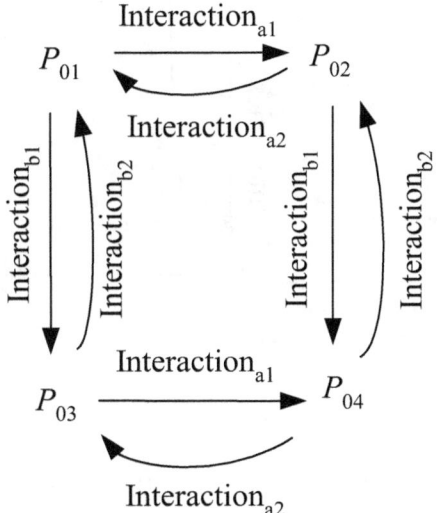

114

We syntactically represent multi-queue SBC process Q_{01} as as FixIFD$_e$||FixIFD$_f$ which equals to "**fix**(X_{05}=Interaction$_{a1}$•Interaction$_{a2}$•λ•X_{05}) || **fix**(X_{06}=Interaction$_{b1}$•Interaction$_{b2}$•λ•X_{06})". The following transition graph shows the semantics of process Q_{01}.

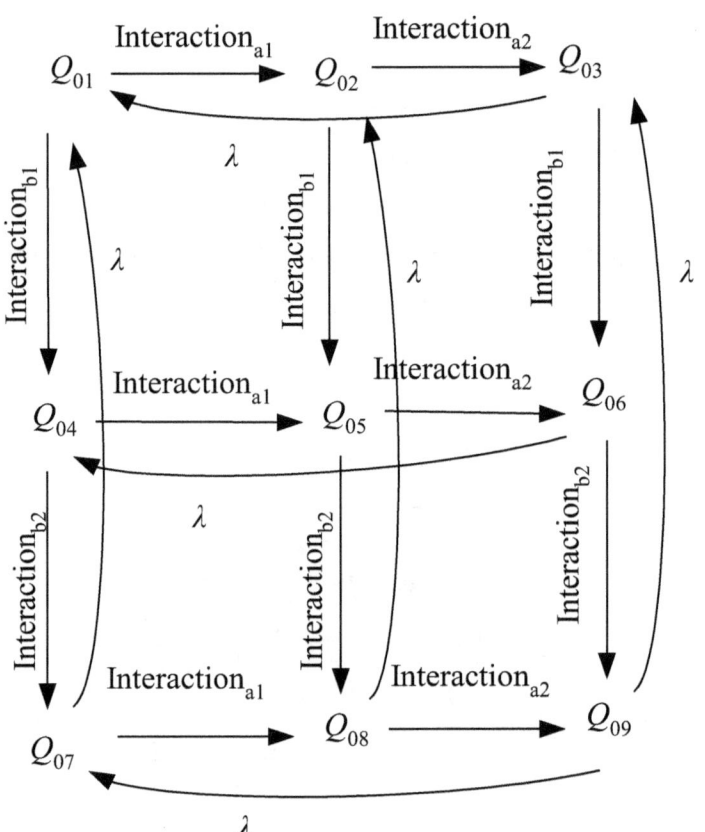

We can easily verify that $S = \{(P_{01}, Q_{01}), (P_{01}, Q_{03}), (P_{01}, Q_{07}), (P_{01}, Q_{09}), (P_{02}, Q_{02}), (P_{02}, Q_{08}), (P_{03}, Q_{04}), (P_{03}, Q_{06}), (P_{04}, Q_{05})\}$ is a bisimulation.

Using the S bisimulation, we then are able to verify that P_{01} and Q_{01} are observation congruent because (1)

$$P_{01} \xrightarrow{\text{Interaction}_{a1}} P_{02}, \text{ then we have } Q_{02} \text{ that } Q_{01} \xRightarrow{\text{Interaction}_{a1}} Q_{02}$$

and $P_{02} \stackrel{\approx}{} Q_{02}$, and (2) $P_{01} \xrightarrow{\text{Interaction}_{b1}} P_{03}$, then we have Q_{04} that

$$Q_{01} \xRightarrow{\text{Interaction}_{b1}} Q_{04} \text{ and } P_{03} \stackrel{\approx}{} Q_{04}, \text{ and (3) } Q_{01} \xrightarrow{\text{Interaction}_{a1}} Q_{02},$$

then we have P_{02} that $P_{01} \xRightarrow{\text{Interaction}_{a1}} P_{02}$ and $P_{02} \stackrel{\approx}{} Q_{02}$, and (4)

$$Q_{01} \xrightarrow{\text{Interaction}_{b1}} Q_{04}, \text{ then we have } P_{03} \text{ that } P_{01} \xRightarrow{\text{Interaction}_{b1}} P_{03}$$

and $P_{03} \stackrel{\approx}{} Q_{04}$.

$P_{01} = Q_{01}$ means that "the concept view of *Kurdi University*" and "the structural composition of the analysis view of *Kurdi University*" are observation congruent.

That is, we have demonstrated that the analysis view of *Kurdi University* is one level down structural decomposition (with observation congruence verification) of the concept view of *Kurdi University*.

Multi-Queue SBC Process of the Design View

We draw the Architecture Hierarchy Diagram of the design view of *Kurdi University* as follows:

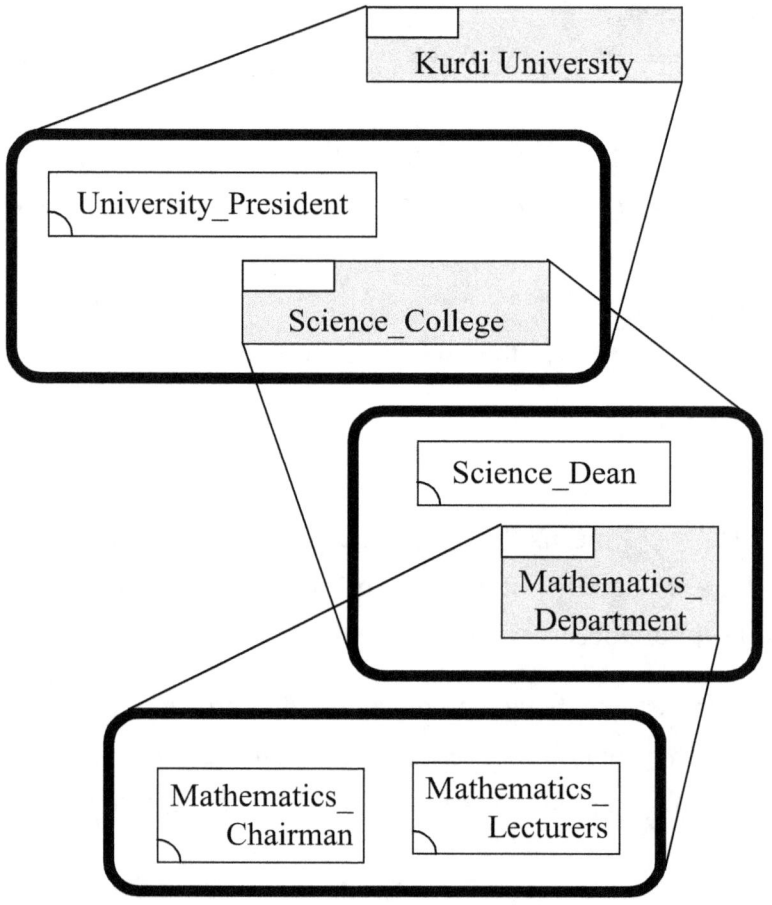

The overall behavior of the design view of *Kurdi University* includes two behaviors: *Study_Calculus_Course* and *Study_Algebra_Course*. Each of them is described by an individual IFD.

An IFD of the *Study_Calculus_Course* behavior is shown below. First, actor *Student* interacts with the *University_President* component through the *University_Teach_Calculus* operation call interaction. Next, component *University_President* interacts with the *Science_Dean* component through the *College_Teach_Calculus* operation call interaction. Continuingly, component *Science_Dean* interacts with the *Mathematics_Chairman* component through the *Department_Teach_Calculus* operation call interaction. Finally, component *Mathematics_Chairman* interacts with the *Mathematics_Lecturers* component through the *Lecturer_Teach_Calculus* operation call interaction.

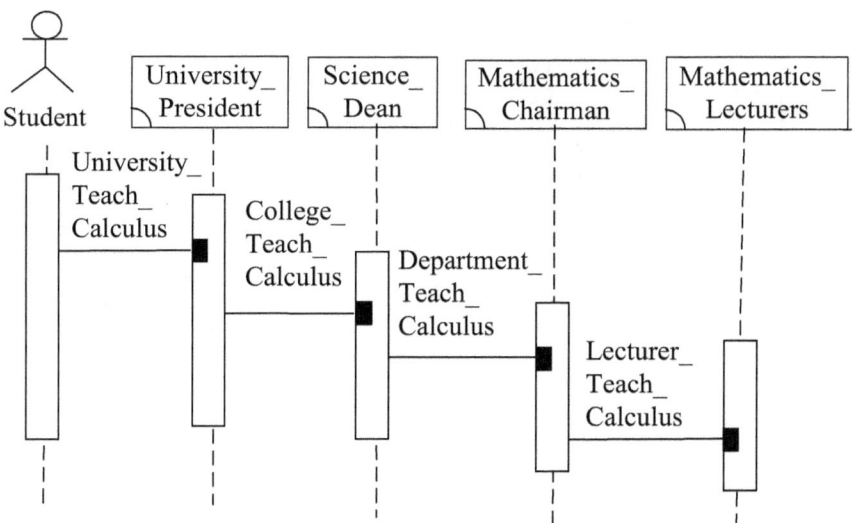

An IFD of the *Study_Algebra_Course* behavior is shown below. First, actor *Student* interacts with the *University_President* component through the *University_Teach_Algebra* operation call interaction. Next, component *University_President* interacts with the *Science_Dean* component through the *College_Teach_Algebra* operation call interaction. Continuingly, component *Science_Dean* interacts with the *Mathematics_Chairman* component through the *Department_Teach_Algebra* operation call interaction. Finally, component *Mathematics_Chairman* interacts with the *Mathematics_Lecturers* component through the *Lecturer_Teach_Algebra* operation call interaction.

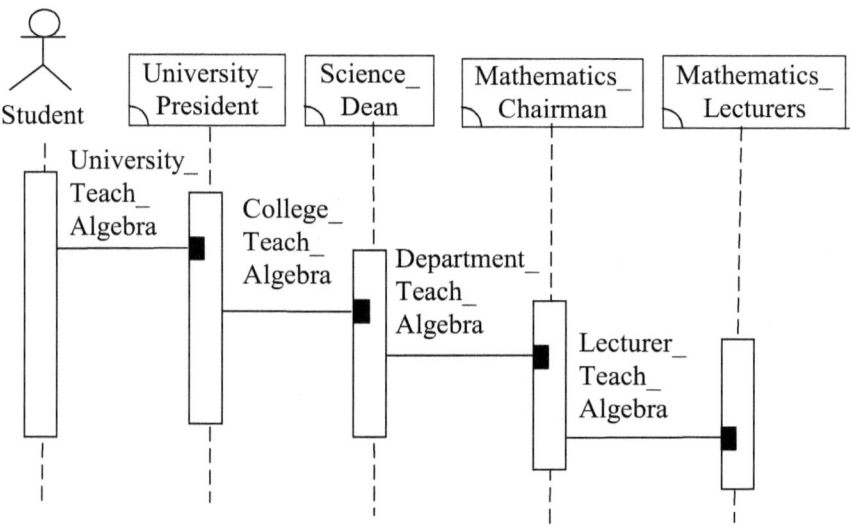

We draw the multi-queue SBC process algebra Backus-Naur Form tree of the design view of *Kurdi University* as follows:

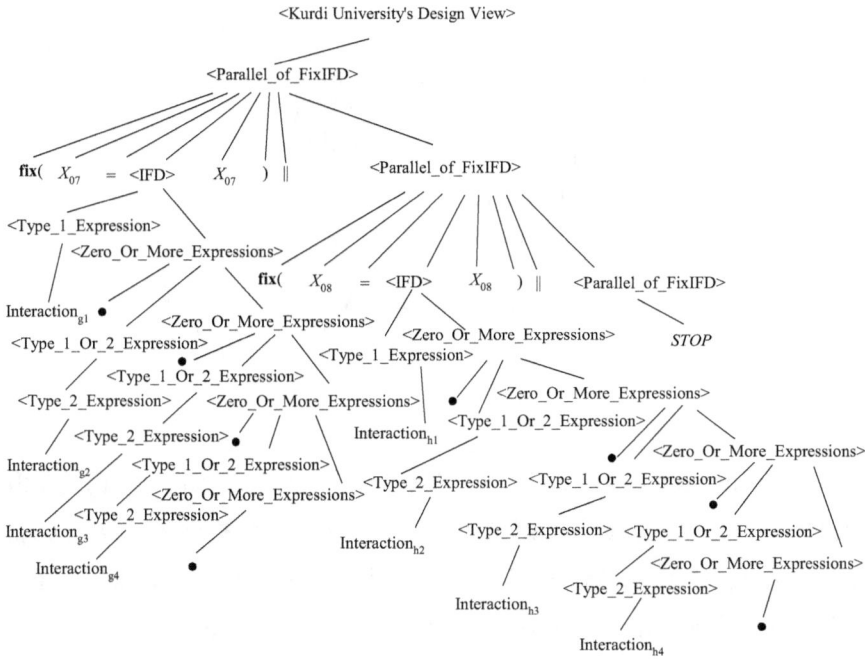

Interaction$_{g1}$ stands for the 1st interaction of the gth interaction flow diagram of the design view of *Kurdi University*. Interaction$_{g1}$ is a type_1 interaction which describes the *Student* actor interacts with the *University_President* component.

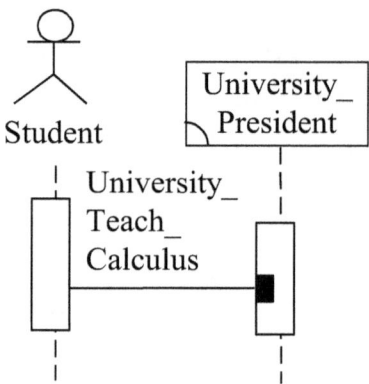

Interaction$_{g2}$ stands for the 2nd interaction of the gth interaction flow diagram of the design view of *Kurdi University*. Interaction$_{g2}$ is a type_2 interaction which describes the *University_President* component interacts with the *Science_Dean* component.

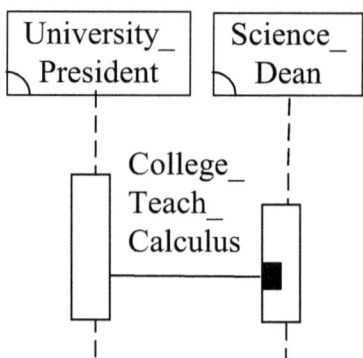

Interaction$_{g3}$ stands for the 3rd interaction of the gth interaction flow diagram of the design view of *Kurdi University*. Interaction$_{g3}$ is a type_2 interaction which describes the *Science_Dean* component interacts with the *Mathematics_Chairman* component.

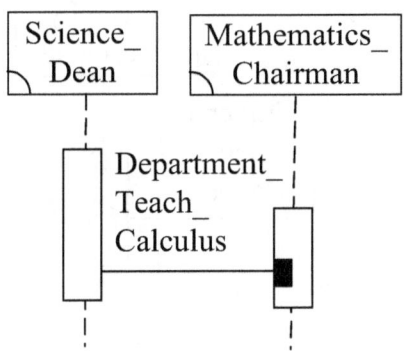

Interaction$_{g4}$ stands for the 4th interaction of the gth interaction flow diagram of the design view of *Kurdi University*. Interaction$_{g4}$ is a type_2 interaction which describes the *Mathematics_Chairman* component interacts with the *Mathematics_Lecturers* component.

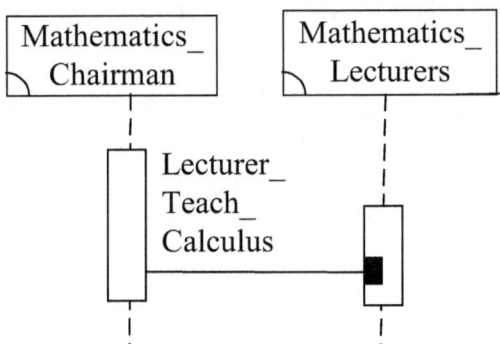

Interaction$_{h1}$ stands for the 1st interaction of the hth interaction flow diagram of the design view of *Kurdi University*. Interaction$_{h1}$ is a type_1 interaction which describes the *Student* actor interacts with the *University_President* component.

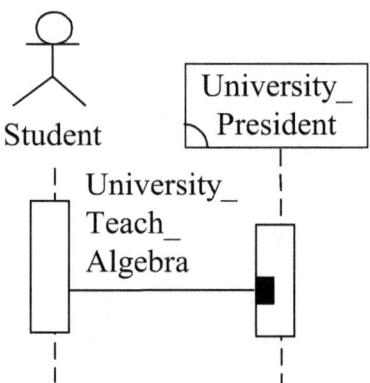

Interaction$_{h2}$ stands for the 2nd interaction of the hth interaction flow diagram of the design view of *Kurdi University*. Interaction$_{h2}$ is a type_2 interaction which describes the *University_President* component interacts with the *Science_Dean* component.

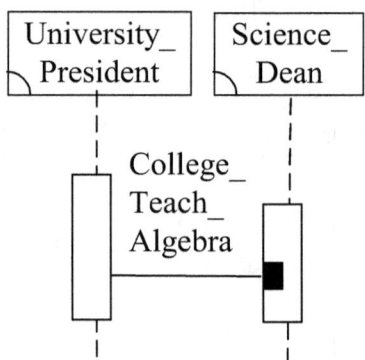

Interaction$_{h3}$ stands for the 3rd interaction of the hth interaction flow diagram of the design view of *Kurdi University*. Interaction$_{h3}$ is a type_2 interaction which describes the *Science_Dean* component interacts with the *Mathematics_Chairman* component.

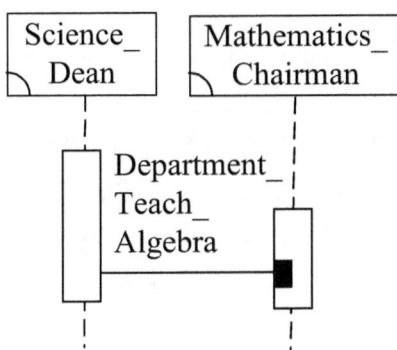

Interaction$_{h4}$ stands for the 4th interaction of the hth interaction flow diagram of the design view of *Kurdi University*. Interaction$_{h4}$ is a type_2 interaction which describes the *Mathematics_Chairman* component interacts with the *Mathematics_Lecturers* component.

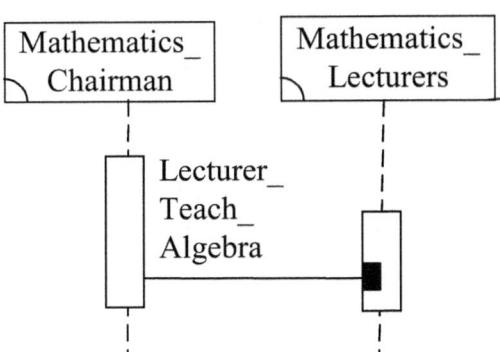

FixIFD$_g$ describes the recursion of the gth interaction flow diagram, i.e. *Study_Calculus_Course* behavior, of the design view of *Kurdi University*. FixIFD$_g$ is syntactically represented as "**fix**(X_{07} = Interaction$_{g1}$●Interaction$_{g2}$●Interaction$_{g3}$●Interaction$_{g4}$●X_{07})".

FixIFD$_g$ $\overset{\text{def}}{=\!=}$

fix(X_{07} = Interaction$_{g1}$● Interaction$_{g2}$●Interaction$_{g3}$●Interaction$_{g4}$●X_{07})

FixIFD$_h$ describes the recursion of the hth interaction flow diagram, i.e. *Study_Algebra_Course* behavior, of the design view of *Kurdi University*. FixIFD$_h$ is syntactically represented as "**fix**(X_{08} = Interaction$_{h1}$●Interaction$_{h2}$●Interaction$_{h3}$●Interaction$_{h4}$●X_{08})".

FixIFD$_h$ $\overset{\text{def}}{=\!=}$

fix(X_{08} = Interaction$_{h1}$● Interaction$_{h2}$●Interaction$_{h3}$●Interaction$_{h4}$●X_{08})

Multi-queue SBC process of the design view of *Kurdi University* is syntactically represented as $FixIFD_g \| FixIFD_h$ which equals to "**fix**$(X_{07}=$ $Interaction_{g1} \bullet Interaction_{g2} \bullet Interaction_{g3} \bullet Interaction_{g4} \bullet X_{07})$ $\|$ **fix**$(X_{08}=$ $Interaction_{h1} \bullet Interaction_{h2} \bullet Interaction_{h3} \bullet Interaction_{h4} \bullet X_{08})$".

Kurdi University's Design View $\overset{\text{def}}{=\joinrel=}$

fix$(X_{07}=Interaction_{g1} \bullet Interaction_{g2} \bullet Interaction_{g3} \bullet Interaction_{g4} \bullet X_{07})$ $\|$
fix$(X_{08}=Interaction_{h1} \bullet Interaction_{h2} \bullet Interaction_{h3} \bullet Interaction_{h4} \bullet X_{08})$

Multi-Queue SBC Process of the Structural Composition of the Design View

Structural composition of the design view of *Kurdi University* means to compose the *Mathematics_Chairman* and *Mathematics_Lecturers* components into the *Mathematics_Department* component. That is, we will rename the *Mathematics_Chairman* component to the *Mathematics_Department* component; we also will rename the *Mathematics_Lecturers* component to the *Mathematics_Department* component.

[Mathematics_Department/Mathematics_Chairman, Mathematics_Department/Mathematics_Lecturers]

We draw the multi-queue SBC process algebra Backus-Naur Form tree of the structural composition of the design view of *Kurdi University* as follows:

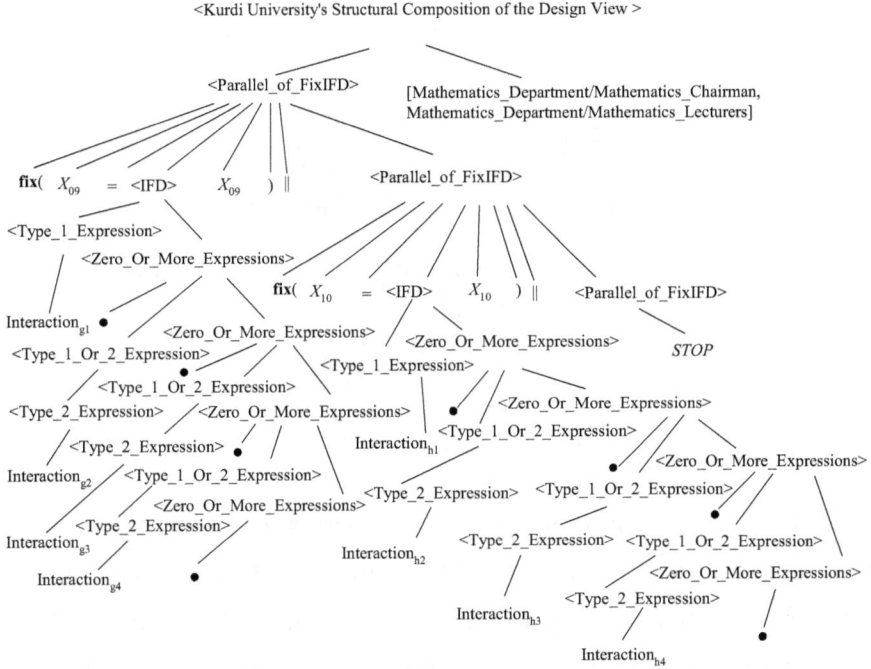

Interaction$_{i1}$=Interaction$_{g1}$[Mathematics_Department/Mathe matics_Chairman,Mathematics_Department/Mathematics_Lecturer s]=Interaction$_{c1}$ stands for the 1st interaction of the ith interaction flow diagram of the structural composition of the design view of *Kurdi University*. Interaction$_{i1}$ is a type_1 interaction which describes the *Student* actor interacts with the *University_President* component.

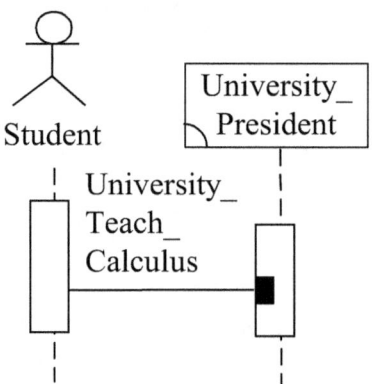

Interaction$_{i2}$=Interaction$_{g2}$[Mathematics_Department/Mathe matics_Chairman,Mathematics_Department/Mathematics_Lecturer s]=Interaction$_{c2}$ stands for the 2nd interaction of the ith interaction flow diagram of the structural composition of the design view of *Kurdi University*. Interaction$_{i2}$ is a type_2 interaction which describes the *University_President* component interacts with the *Science_Dean* component.

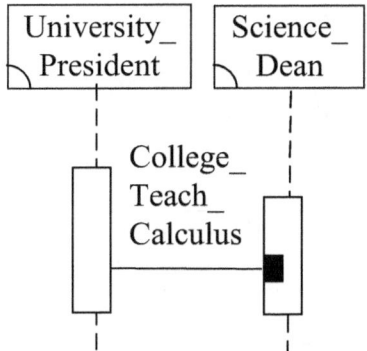

Interaction$_{i3}$=Interaction$_{g3}$[Mathematics_Department/Mathe matics_Chairman,Mathematics_Department/Mathematics_Lecturer s]=Interaction$_{c3}$ stands for the 3rd interaction of the ith interaction flow diagram of the structural composition of the design view of *Kurdi University*. Interaction$_{i3}$ is a type_2 interaction which describes the *Science_Dean* component interacts with the *Mathematics_Department* component.

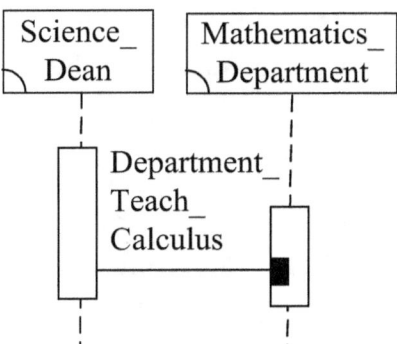

Interaction$_{i4}$=Interaction$_{g4}$[Mathematics_Department/Mathe matics_Chairman,Mathematics_Department/Mathematics_Lecturer s] stands for the 4th interaction of the ith interaction flow diagram of the structural composition of the design view of *Kurdi University*. Interaction$_{i4}$ is an internal interaction (i.e. λ) inside the *Mathematics_Department* component.

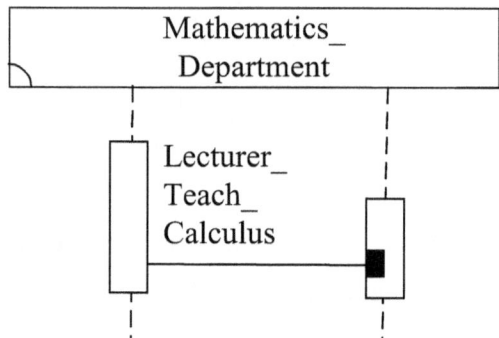

Interaction$_{j1}$=Interaction$_{h1}$[Mathematics_Department/Mathematics_Chairman,Mathematics_Department/Mathematics_Lecturers]=Interaction$_{d1}$ stands for the 1st interaction of the jth interaction flow diagram of the structural composition of the design view of *Kurdi University*. Interaction$_{j1}$ is a type_1 interaction which describes the *Student* actor interacts with the *University_President* component.

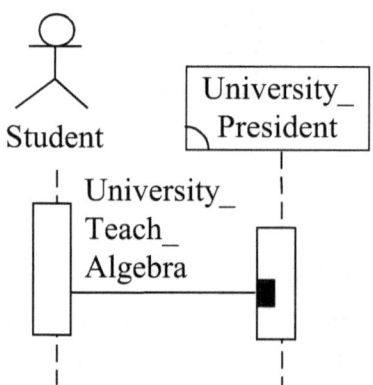

Interaction$_{j2}$=Interaction$_{h2}$[Mathematics_Department/Mathe matics_Chairman,Mathematics_Department/Mathematics_Lecturer s]=Interaction$_{d2}$ stands for the 2nd interaction of the jth interaction flow diagram of the structural composition of the design view of *Kurdi University*. Interaction$_{j2}$ is a type_2 interaction which describes the *University_President* component interacts with the *Science_Dean* component.

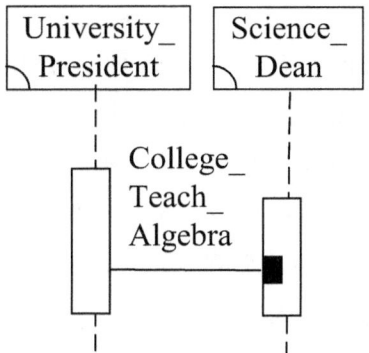

Interaction$_{j3}$=Interaction$_{h3}$[Mathematics_Department/Mathe matics_Chairman,Mathematics_Department/Mathematics_Lecturer s]=Interaction$_{d3}$ stands for the 3rd interaction of the jth interaction flow diagram of the structural composition of the design view of *Kurdi University*. Interaction$_{j3}$ is a type_2 interaction which describes the *Science_Dean* component interacts with the *Mathematics_Department* component.

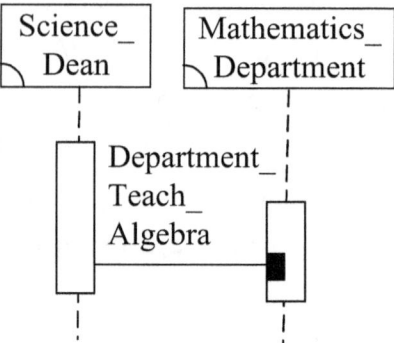

Interaction$_{j4}$=Interaction$_{h4}$[Mathematics_Department/Mathe
matics_Chairman,Mathematics_Department/Mathematics_Lecturer
s] stands for the 4th interaction of the jth interaction flow diagram
of the structural composition of the design view of *Kurdi
University*. Interaction$_{j4}$ is an internal interaction (i.e. λ) inside the
Mathematics_Department component.

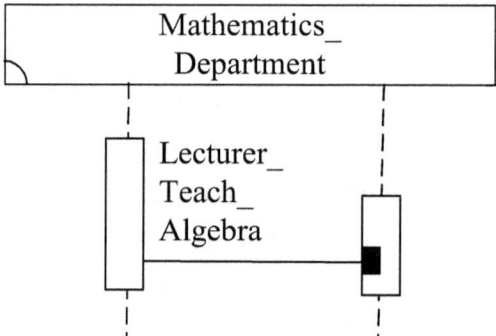

FixIFD$_i$ describes the recursion of the ith interaction flow
diagram, i.e. *Study_Calculus_Course* behavior, of the structural
composition of the design view of *Kurdi University*. FixIFD$_i$ is
syntactically   represented   as   "**fix**(X_{09} =
Interaction$_{c1}$•Interaction$_{c2}$•Interaction$_{c3}$•λ•X_{09})".

$$\text{FixIFD}_i \stackrel{\text{def}}{==}$$

$$\mathbf{fix}(X_{09} = \text{Interaction}_{c1} \bullet \text{Interaction}_{c2} \bullet \text{Interaction}_{c3} \bullet \lambda \bullet X_{09})$$

FixIFD$_j$ describes the recursion of the jth interaction flow diagram, i.e. *Study_Algebra_Course* behavior, of the structural composition of the design view of *Kurdi University*. FixIFD$_j$ is syntactically represented as "$\mathbf{fix}(X_{10}$ = Interaction$_{d1}$•Interaction$_{d2}$•Interaction$_{d3}$•λ•$X_{10})$".

$$\text{FixIFD}_j \stackrel{\text{def}}{==}$$

$$\mathbf{fix}(X_{10} = \text{Interaction}_{d1} \bullet \text{Interaction}_{d2} \bullet \text{Interaction}_{d3} \bullet \lambda \bullet X_{10})$$

Multi-queue SBC process of the structural composition of the design view of *Kurdi University* is syntactically represented as FixIFD$_i$‖FixIFD$_j$ which equals to "$\mathbf{fix}(X_{09}$ = Interaction$_{c1}$•Interaction$_{c2}$•Interaction$_{c3}$•λ•X_{09}) ‖ $\mathbf{fix}(X_{10}$ = Interaction$_{d1}$•Interaction$_{d2}$•Interaction$_{d3}$•λ•$X_{10})$".

Kurdi University's Structural Composition of the Design View $\stackrel{\text{def}}{==}$

$$\mathbf{fix}(X_{09}=\text{Interaction}_{c1} \bullet \text{Interaction}_{c2} \bullet \text{Interaction}_{c3} \bullet \lambda \bullet X_{09}) \; \|$$
$$\mathbf{fix}(X_{10}=\text{Interaction}_{d1} \bullet \text{Interaction}_{d2} \bullet \text{Interaction}_{d3} \bullet \lambda \bullet X_{10})$$

Observation Congruence of "the Analysis View" and "the Structural Composition of the Design View"

We syntactically represent multi-queue SBC process P_{01} as $FixIFD_c \| FixIFD_d$ which equals to "**fix**$(X_{03} = Interaction_{c1} \bullet Interaction_{c2} \bullet Interaction_{c3} \bullet X_{03})$ $\|$ **fix**$(X_{04} = Interaction_{d1} \bullet Interaction_{d2} \bullet Interaction_{d3} \bullet X_{04})$". The following transition graph shows the semantics of process P_{01}.

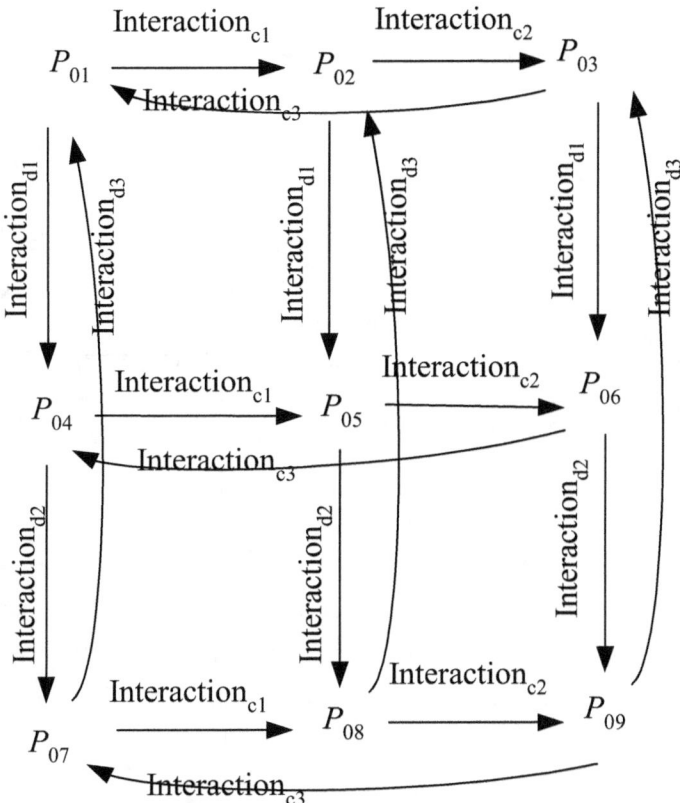

We syntactically represent process Q_{01} as FixIFD$_i$‖FixIFD$_j$ which equals to "**fix**(X_{09} = Interaction$_{c1}$●Interaction$_{c2}$●Interaction$_{c3}$●λ●X_{09}) ‖ **fix**(X_{10} = Interaction$_{d1}$●Interaction$_{d2}$●Interaction$_{d3}$●λ●X_{10})". The following transition graph shows the semantics of process Q_{01}.

136

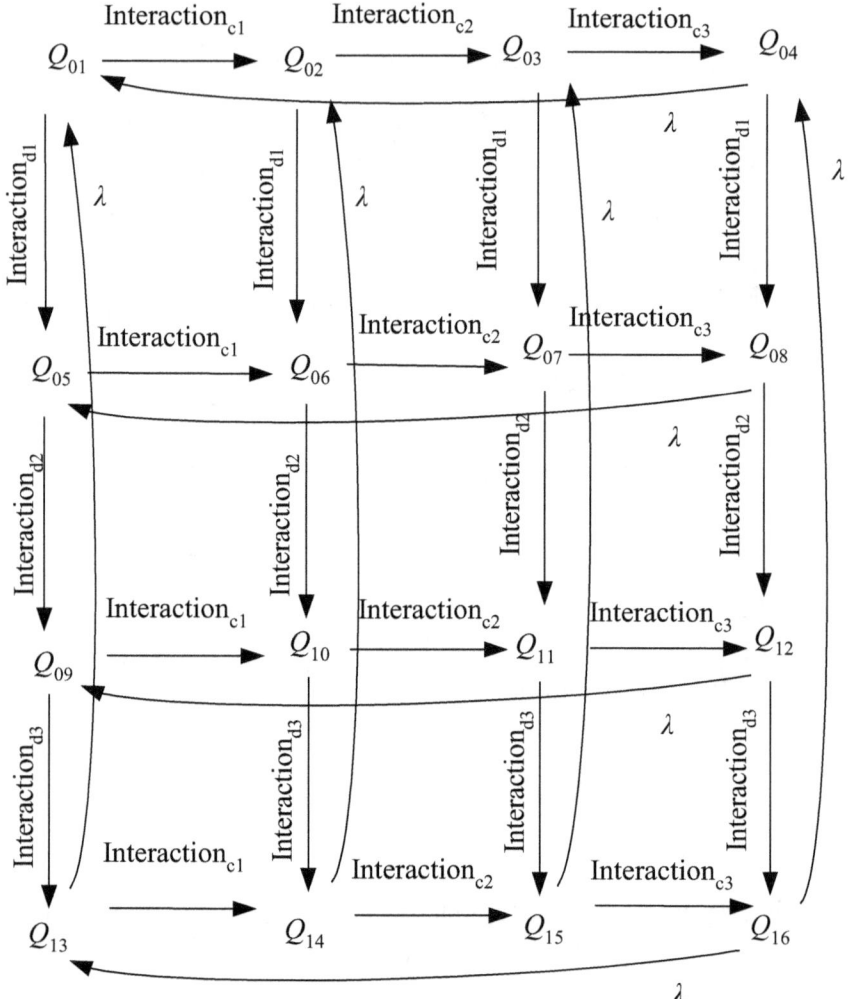

We can easily verify that $S = \{(P_{01}, Q_{01}), (P_{01}, Q_{04}), (P_{01}, Q_{13}), (P_{01}, Q_{16}), (P_{02}, Q_{02}), (P_{02}, Q_{14}), (P_{03}, Q_{03}), (P_{03}, Q_{15}), (P_{04}, Q_{05}), (P_{04}, Q_{08}), (P_{05}, Q_{06}), (P_{06}, Q_{07}), (P_{07}, Q_{09}), (P_{07}, Q_{12}), (P_{08}, Q_{10}), (P_{09}, Q_{11})\}$ is a bisimulation.

Using the S bisimulation, we then are able to verify that P_{01}

and Q_{01} are observation congruent because (1) $P_{01} \xrightarrow{\text{Interaction}_{c1}} P_{02}$, then we have Q_{02} that $Q_{01} \xRightarrow{\text{Interaction}_{c1}} Q_{02}$ and $P_{02} \overset{\sim}{\thicksim} Q_{02}$, and (2) $P_{01} \xrightarrow{\text{Interaction}_{d1}} P_{04}$, then we have Q_{05} that $Q_{01} \xRightarrow{\text{Interaction}_{d1}} Q_{05}$ and $P_{04} \overset{\sim}{\thicksim} Q_{05}$, and (3) $Q_{01} \xrightarrow{\text{Interaction}_{c1}} Q_{02}$, then we have P_{02} that $P_{01} \xRightarrow{\text{Interaction}_{c1}} P_{02}$ and $P_{02} \overset{\sim}{\thicksim} Q_{02}$, and (4) $Q_{01} \xrightarrow{\text{Interaction}_{d1}} Q_{05}$, then we have P_{04} that $P_{01} \xRightarrow{\text{Interaction}_{d1}} P_{04}$ and $P_{04} \overset{\sim}{\thicksim} Q_{05}$.

$P_{01} = Q_{01}$ means that "the analysis view of *Kurdi University*" and "the structural composition of the design view of *Kurdi University*" are observation congruent.

That is, we have demonstrated that the design view of *Kurdi University* is one level down structural decomposition (with observation congruence verification) of the analysis view of *Kurdi University*.

Multi-Queue SBC Process of the Implementation View

We draw the Architecture Hierarchy Diagram (AHD) of the implementation view of *Kurdi University* as follows:

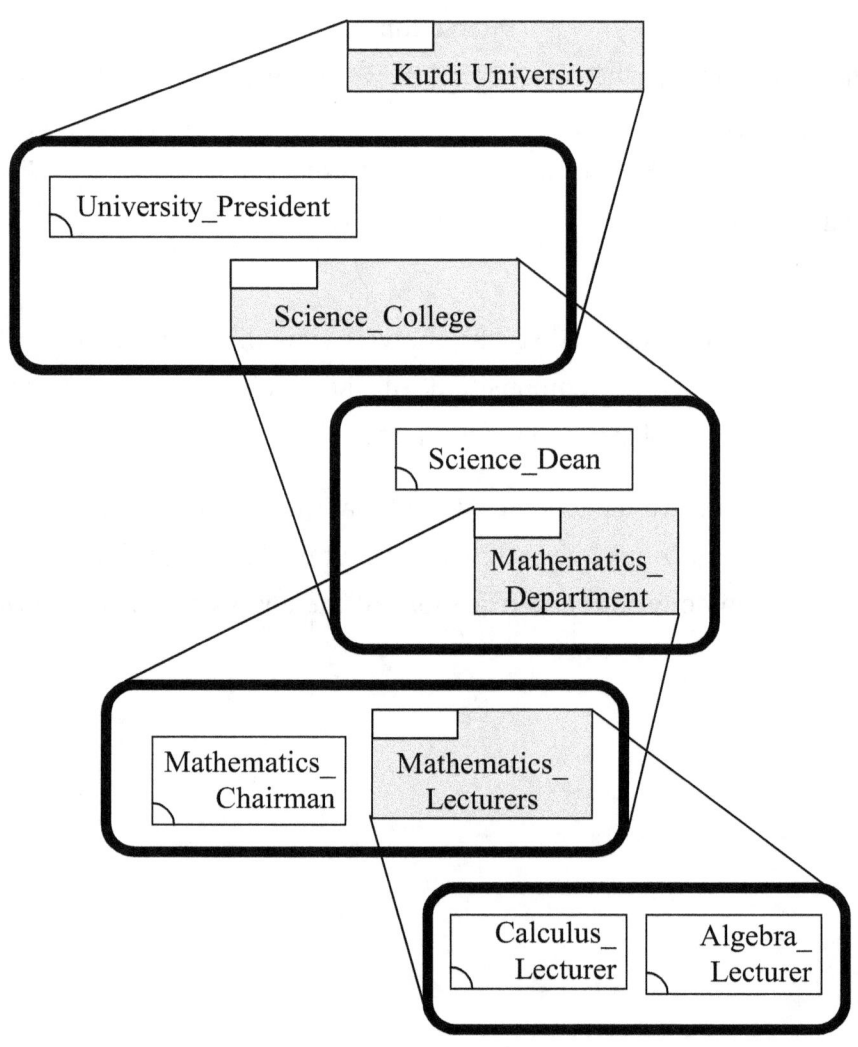

The overall behavior of the implementation view of *Kurdi University* includes two behaviors: *Study_Calculus_Course* and *Study_Algebra_Course*. Each of them is described by an individual IFD.

An IFD of the *Study_Calculus_Course* behavior is shown below. First, actor *Student* interacts with the *University_President* component through the *University_Teach_Calculus* operation call interaction. Next, component *University_President* interacts with the *Science_Dean* component through the *College_Teach_Calculus* operation call interaction. Continuingly, component *Science_Dean* interacts with the *Mathematics_Chairman* component through the *Department_Teach_Calculus* operation call interaction. Finally, component *Mathematics_Chairman* interacts with the *Calculus_Lecturer* component through the *Lecturer_Teach_Calculus* operation call interaction.

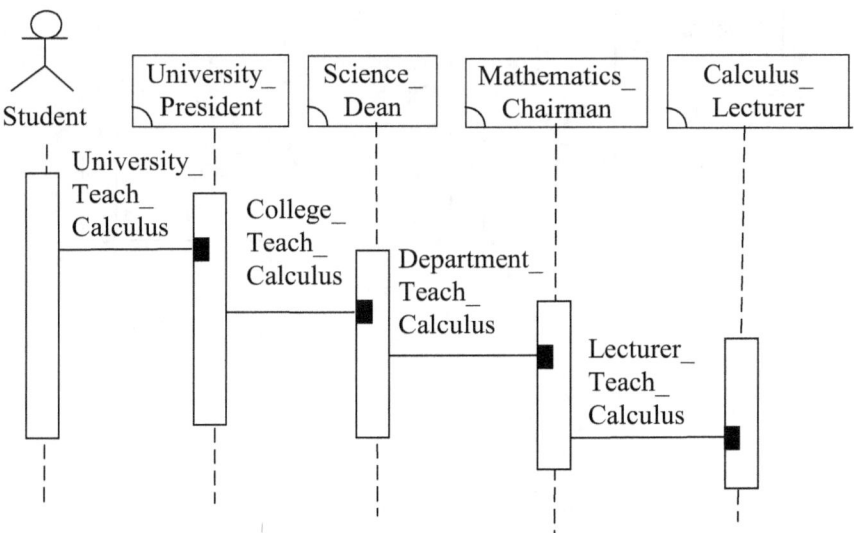

An IFD of the *Study_Algebra_Course* behavior is shown below. First, actor *Student* interacts with the *University_President* component through the *University_Teach_Algebra* operation call interaction. Next, component *University_President* interacts with the *Science_Dean* component through the *College_Teach_Algebra* operation call interaction. Continuingly, component *Science_Dean* interacts with the *Mathematics_Chairman* component through the *Department_Teach_Algebra* operation call interaction. Finally, component *Mathematics_Chairman* interacts with the *Algebra_Lecturer* component through the *Lecturer_Teach_Algebra* operation call interaction.

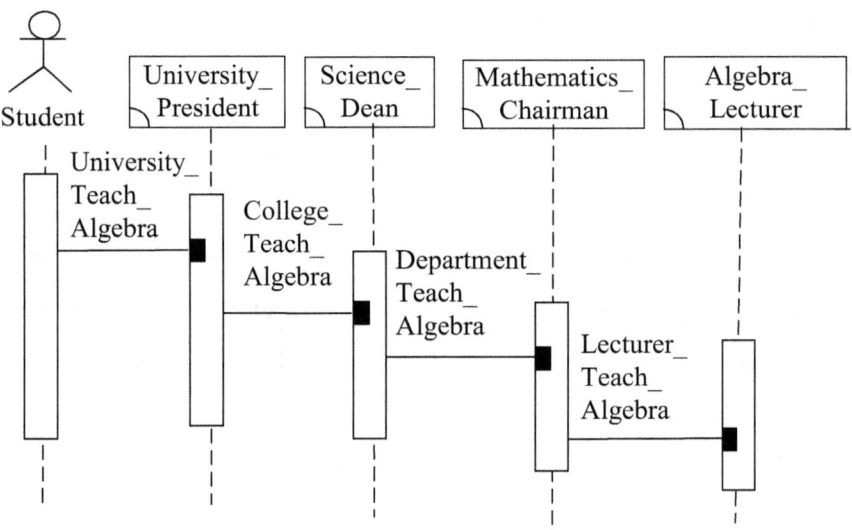

We draw the multi-queue SBC process algebra Backus-Naur Form tree of the implementation view of *Kurdi University* as follows:

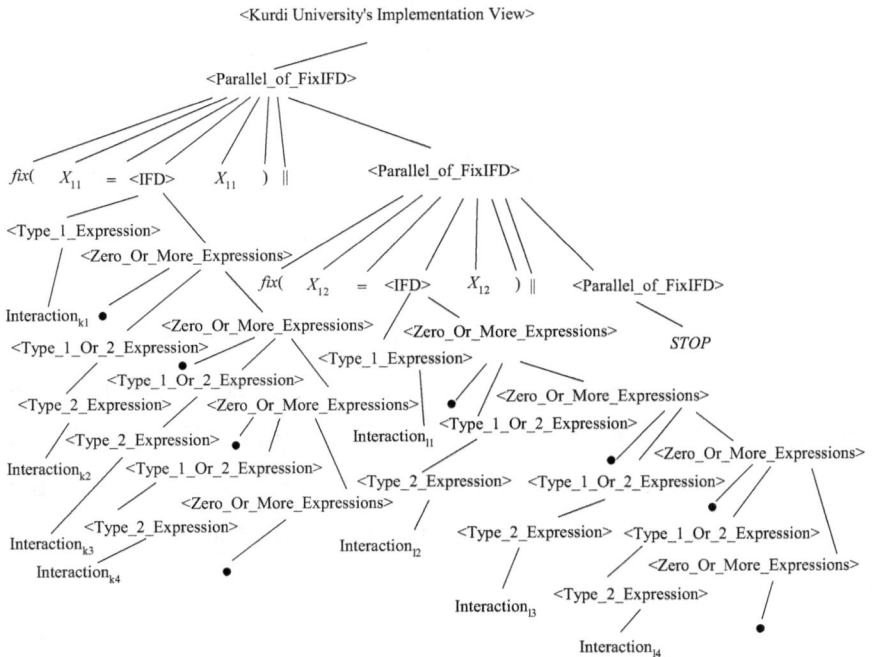

Interaction$_{k1}$ stands for the 1st interaction of the kth interaction flow diagram of the implementation view of *Kurdi University*. Interaction$_{k1}$ is a type_1 interaction which describes the *Student* actor interacts with the *University_President* component.

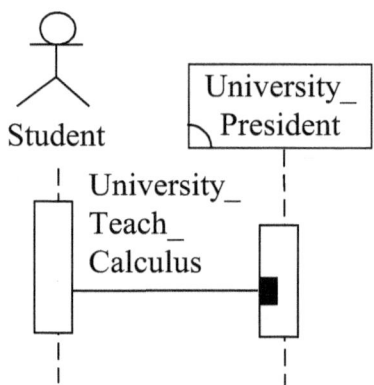

Interaction$_{k2}$ stands for the 2nd interaction of the kth interaction flow diagram of the implementation view of *Kurdi University*. Interaction$_{k2}$ is a type_2 interaction which describes the *University_President* component interacts with the *Science_Dean* component.

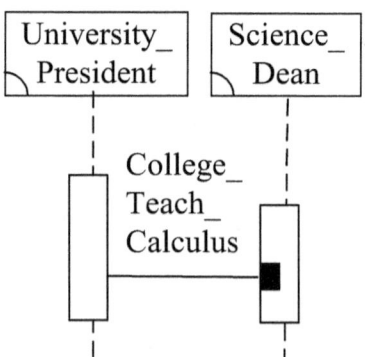

Interaction$_{k3}$ stands for the 3rd interaction of the kth interaction flow diagram of the implementation view of *Kurdi University*. Interaction$_{k3}$ is a type_2 interaction which describes the *Science_Dean* component interacts with the *Mathematics_Chairman* component.

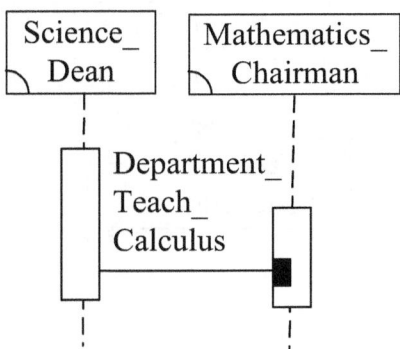

Interaction$_{k4}$ stands for the 4th interaction of the kth interaction flow diagram of the implementation view of *Kurdi University*. Interaction$_{k4}$ is a type_2 interaction which describes the *Mathematics_Chairman* component interacts with the *Calculus_Lecturer* component.

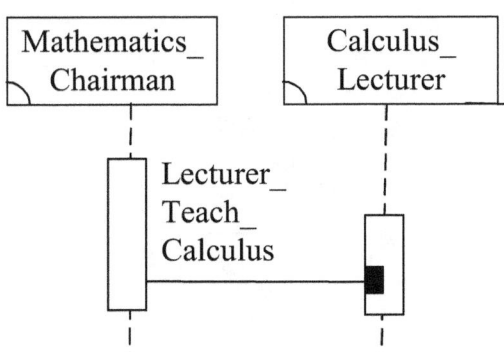

144

Interaction$_{l1}$ stands for the 1st interaction of the lth interaction flow diagram of the implementation view of *Kurdi University*. Interaction$_{l1}$ is a type_1 interaction which describes the *Student* actor interacts with the *University_President* component.

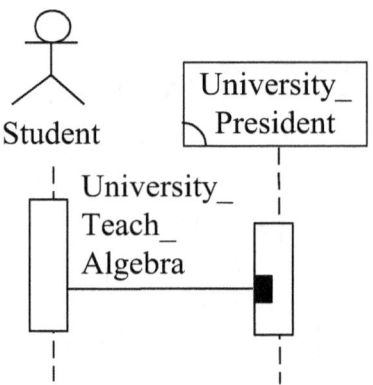

Interaction$_{l2}$ stands for the 2nd interaction of the lth interaction flow diagram of the implementation view of *Kurdi University*. Interaction$_{l2}$ is a type_2 interaction which describes the *University_President* component interacts with the *Science_Dean* component.

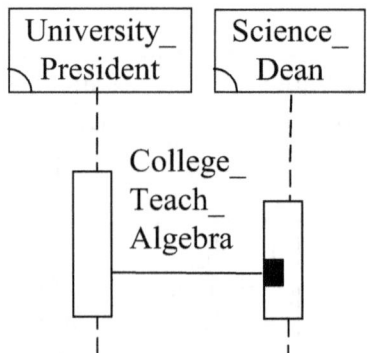

Interaction$_{13}$ stands for the 3rd interaction of the lth interaction flow diagram of the implementation view of *Kurdi University*. Interaction$_{13}$ is a type_2 interaction which describes the *Science_Dean* component interacts with the *Mathematics_Chairman* component.

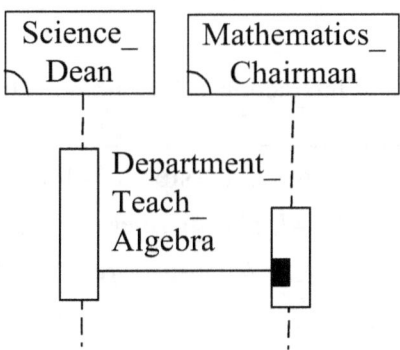

Interaction$_{14}$ stands for the 4th interaction of the lth interaction flow diagram of the implementation view of *Kurdi University*. Interaction$_{14}$ is a type_2 interaction which describes the *Mathematics_Chairman* component interacts with the *Algebra_Lecturer* component.

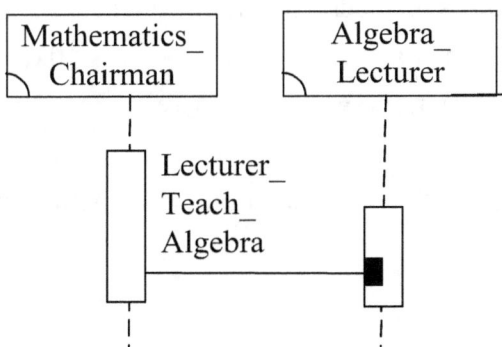

FixIFD$_k$ describes the recursion of the kth interaction flow diagram, i.e. *Study_Calculus_Course* behavior, of the implementation view of *Kurdi University*. FixIFD$_k$ is syntactically represented as "**fix**(X_{11} = Interaction$_{k1}$•Interaction$_{k2}$•Interaction$_{k3}$•Interaction$_{k4}$•X_{11})".

$$\text{FixIFD}_k \;\overset{\text{def}}{=\!=}$$

$$\mathbf{fix}(X_{11} = \text{Interaction}_{k1} \bullet \text{Interaction}_{k2} \bullet \text{Interaction}_{k3} \bullet \text{Interaction}_{k4} \bullet X_{11})$$

FixIFD$_l$ describes the recursion of the lth interaction flow diagram, i.e. *Study_Algebra_Course* behavior, of the implementation view of *Kurdi University*. FixIFD$_l$ is syntactically represented as "**fix**(X_{12} = Interaction$_{11}$•Interaction$_{12}$•Interaction$_{13}$•Interaction$_{14}$•X_{12})".

$$\text{FixIFD}_l \;\overset{\text{def}}{=\!=}$$

$$\mathbf{fix}(X_{12} = \text{Interaction}_{11} \bullet \text{Interaction}_{12} \bullet \text{Interaction}_{13} \bullet \text{Interaction}_{14} \bullet X_{12})$$

Multi-queue SBC process of the implementation view of *Kurdi University* is syntactically represented as FixIFD$_k$‖FixIFD$_l$ which equals to "**fix**(X_{11} = Interaction$_{k1}$•Interaction$_{k2}$•Interaction$_{k3}$•Interaction$_{k4}$•X_{11}) ‖ **fix**(X_{12} = Interaction$_{11}$•Interaction$_{12}$•Interaction$_{13}$•Interaction$_{14}$•X_{12})".

Kurdi University's Implementation View $\overset{\text{def}}{=\!=}$

$\mathbf{fix}(X_{11}=\text{Interaction}_{k1}\bullet\ \text{Interaction}_{k2}\bullet\ \text{Interaction}_{k3}\bullet\ \text{Interaction}_{k4}\bullet\ X_{11})\ \parallel$
$\mathbf{fix}(X_{12}=\text{Interaction}_{l1}\bullet\ \text{Interaction}_{l2}\bullet\ \text{Interaction}_{l3}\bullet\ \text{Interaction}_{l4}\bullet\ X_{12})$

Multi-Queue SBC Process of the Structural Composition of the Implementation View

Structural composition of the implementation view of *Kurdi University* means to compose the *Calculus_Lecturer* and *Algebra_Lecturer* components into the *Mathematics_Lecturers* component. That is, we will rename the *Calculus_Lecturer* component to the *Mathematics_Lecturers* component; we also will rename the *Algebra_Lecturer* component to the *Mathematics_Lecturers* component.

[Mathematics_Lecturers/Calculus_Lecturer,
Mathematics_Lecturers/Algebra_Lecturer]

149

We draw the Backus-Naur Form tree of the structural composition of the implementation view of *Kurdi University* as follows:

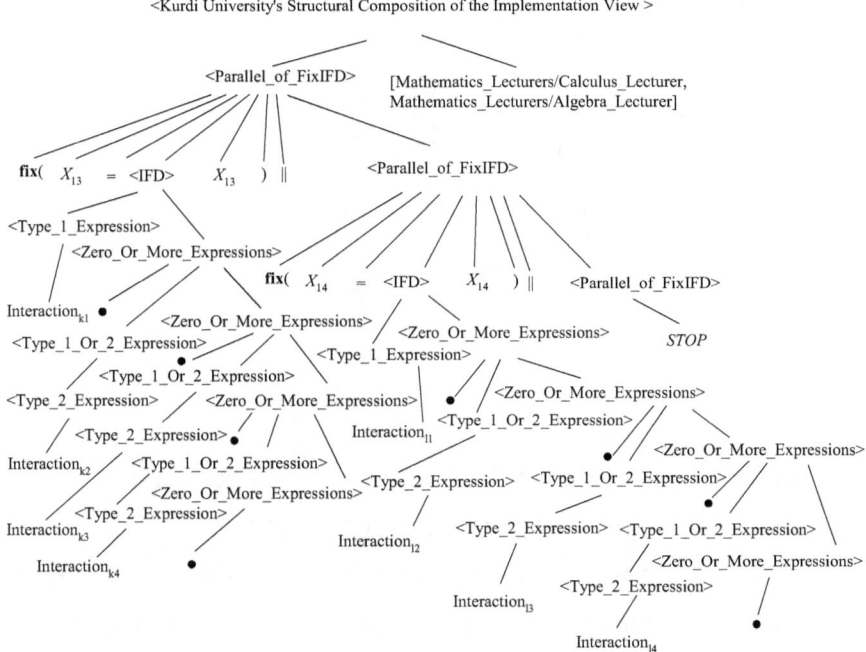

Interaction$_{m1}$=Interaction$_{k1}$[Mathematics_Lecturers/Calculus _Lecturer,Mathematics_Lecturers/Algebra_Lecturer]=Interaction$_{g1}$ stands for the 1st interaction of the mth interaction flow diagram of the structural composition of the implementation view of *Kurdi University*. Interaction$_{m1}$ is a type_1 interaction which describes the *Student* actor interacts with the *University_President* component.

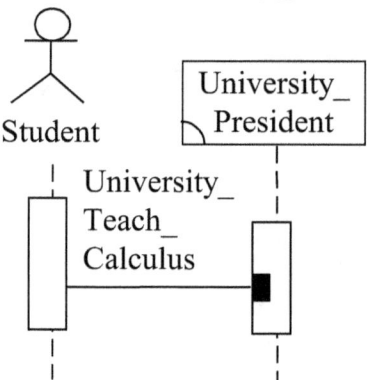

Interaction$_{m2}$=Interaction$_{k2}$[Mathematics_Lecturers/Calculus _Lecturer,Mathematics_Lecturers/Algebra_Lecturer]=Interaction$_{g2}$ stands for the 2nd interaction of the mth interaction flow diagram of the structural composition of the implementation view of *Kurdi University*. Interaction$_{m2}$ is a type_2 interaction which describes the *University_President* component interacts with the *Science_Dean* component.

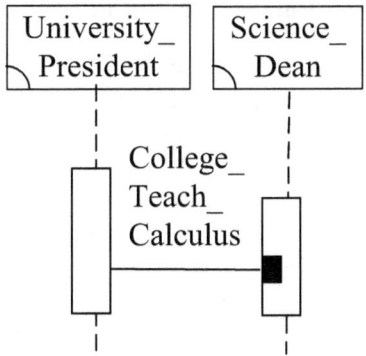

Interaction$_{m3}$=Interaction$_{k3}$[Mathematics_Lecturers/Calculus _Lecturer,Mathematics_Lecturers/Algebra_Lecturer]=Interaction$_{g3}$ stands for the 3rd interaction of the mth interaction flow diagram of the structural composition of the implementation view of *Kurdi University*. Interaction$_{m3}$ is a type_2 interaction which describes the *Science_Dean* component interacts with the *Mathematics_Chairman* component.

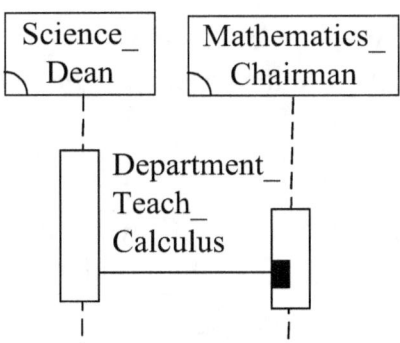

Interaction$_{m4}$=Interaction$_{k4}$[Mathematics_Lecturers/Calculus _Lecturer,Mathematics_Lecturers/Algebra_Lecturer]=Interaction$_{g4}$ stands for the 4th interaction of the mth interaction flow diagram of the structural composition of the implementation view of *Kurdi University*. Interaction$_{m4}$ is a type_2 interaction which describes the *Mathematics_Chairman* component interacts with the *Mathematics_Lecturers* component.

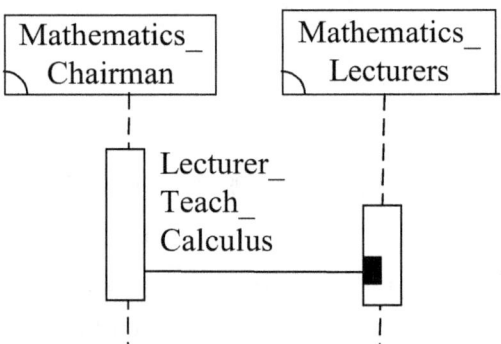

Interaction$_{n1}$=Interaction$_{11}$[Mathematics_Lecturers/Calculus_ Lecturer,Mathematics_Lecturers/Algebra_Lecturer]=Interaction$_{h1}$ stands for the 1st interaction of the nth interaction flow diagram of the structural composition of the implementation view of *Kurdi University*. Interaction$_{n1}$ is a type_1 interaction which describes the *Student* actor interacts with the *University_President* component.

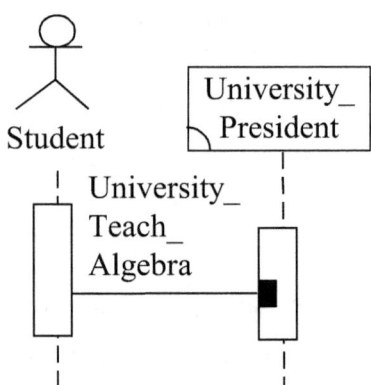

Interaction$_{n2}$=Interaction$_{12}$[Mathematics_Lecturers/Calculus_
Lecturer,Mathematics_Lecturers/Algebra_Lecturer]=Interaction$_{h2}$
stands for the 2nd interaction of the nth interaction flow diagram of
the structural composition of the implementation view of *Kurdi
University*. Interaction$_{n2}$ is a type_2 interaction which describes the
University_President component interacts with the *Science_Dean*
component.

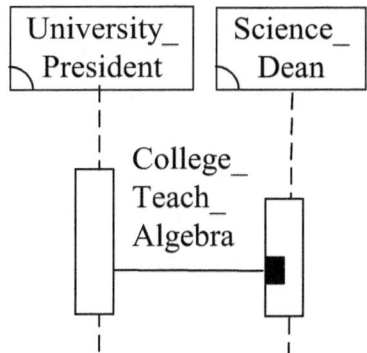

Interaction$_{n3}$=Interaction$_{13}$[Mathematics_Lecturers/Calculus_
Lecturer,Mathematics_Lecturers/Algebra_Lecturer]=Interaction$_{h3}$
stands for the 3rd interaction of the nth interaction flow diagram of
the structural composition of the implementation view of *Kurdi
University*. Interaction$_{n3}$ is a type_2 interaction which describes the
Science_Dean   component   interacts   with   the
Mathematics_Chairman component.

154

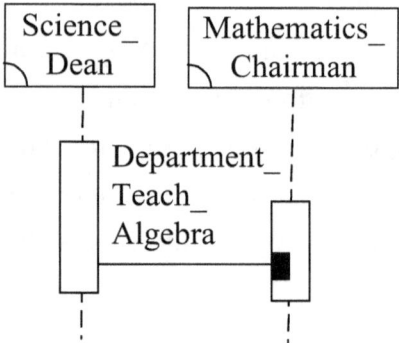

Interaction$_{n4}$=Interaction$_{l4}$[Mathematics_Lecturers/Calculus_
Lecturer,Mathematics_Lecturers/Algebra_Lecturer]=Interaction$_{h4}$
stands for the 4th interaction of the nth interaction flow diagram of
the structural composition of the implementation view of *Kurdi
University*. Interaction$_{n4}$ is a type_2 interaction which describes the
Mathematics_Chairman component interacts with the
Mathematics_Lecturers component.

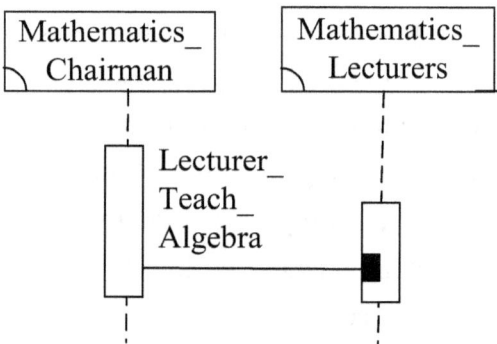

FixIFD$_m$ describes the recursion of the mth interaction flow diagram, i.e. *Study_Calculus_Course* behavior, of the structural composition of the implementation view of *Kurdi University*. FixIFD$_m$ is syntactically represented as "$\mathbf{fix}(X_{13} = $ Interaction$_{g1} \bullet$Interaction$_{g2} \bullet$Interaction$_{g3} \bullet$Interaction$_{g4} \bullet X_{13})$".

$$\text{FixIFD}_m \stackrel{\text{def}}{=\!=}$$

$$\mathbf{fix}(X_{13} = \text{Interaction}_{g1} \bullet \text{Interaction}_{g2} \bullet \text{Interaction}_{g3} \bullet \text{Interaction}_{g4} \bullet X_{13})$$

FixIFD$_n$ describes the recursion of the nth interaction flow diagram, i.e. *Study_Algebra_Course* behavior, of the structural composition of the implementation view of *Kurdi University*. FixIFD$_n$ is syntactically represented as "$\mathbf{fix}(X_{14} = $ Interaction$_{h1} \bullet$Interaction$_{h2} \bullet$Interaction$_{h3} \bullet$Interaction$_{h4} \bullet X_{14})$".

$$\text{FixIFD}_n \stackrel{\text{def}}{=\!=}$$

$$\mathbf{fix}(X_{14} = \text{Interaction}_{h1} \bullet \text{Interaction}_{h2} \bullet \text{Interaction}_{h3} \bullet \text{Interaction}_{h4} \bullet X_{14})$$

156

Multi-queue SBC process of the structural composition of the implementation view of *Kurdi University* is syntactically represented as FixIFD$_m$‖FixIFD$_n$ which equals to "**fix**($X_{13}=$ Interaction$_{g1}$●Interaction$_{g2}$●Interaction$_{g3}$●Interaction$_{g4}$●X_{13})   ‖ **fix**(X_{14} = Interaction$_{h1}$●Interaction$_{h2}$●Interaction$_{h3}$●Interaction$_{h4}$●X_{14})".

Kurdi University's Structural Composition of the Implementation View $\overset{def}{=\!=}$

$fix(X_{13}=$Interaction$_{g1}$ ● Interaction$_{g2}$ ● Interaction$_{g3}$●Interaction$_{g4}$●X_{13}) ‖
$fix(X_{14}=$Interaction$_{h1}$ ● Interaction$_{h2}$ ● Interaction$_{h3}$● Interaction$_{h4}$● X_{14})

Observation Congruence of "the Design View" and "the Structural Composition of the Implementation View"

We syntactically represent multi-queue SBC process P_{01} as FixIFD$_g$||FixIFD$_h$ which equals to "**fix**(X_{07}= Interaction$_{g1}$•Interaction$_{g2}$•Interaction$_{g3}$•Interaction$_{g4}$•X_{07}) || **fix**(X_{08}= Interaction$_{h1}$•Interaction$_{h2}$•Interaction$_{h3}$•Interaction$_{h4}$•X_{08})". The following transition graph shows the semantics of process P_{01}.

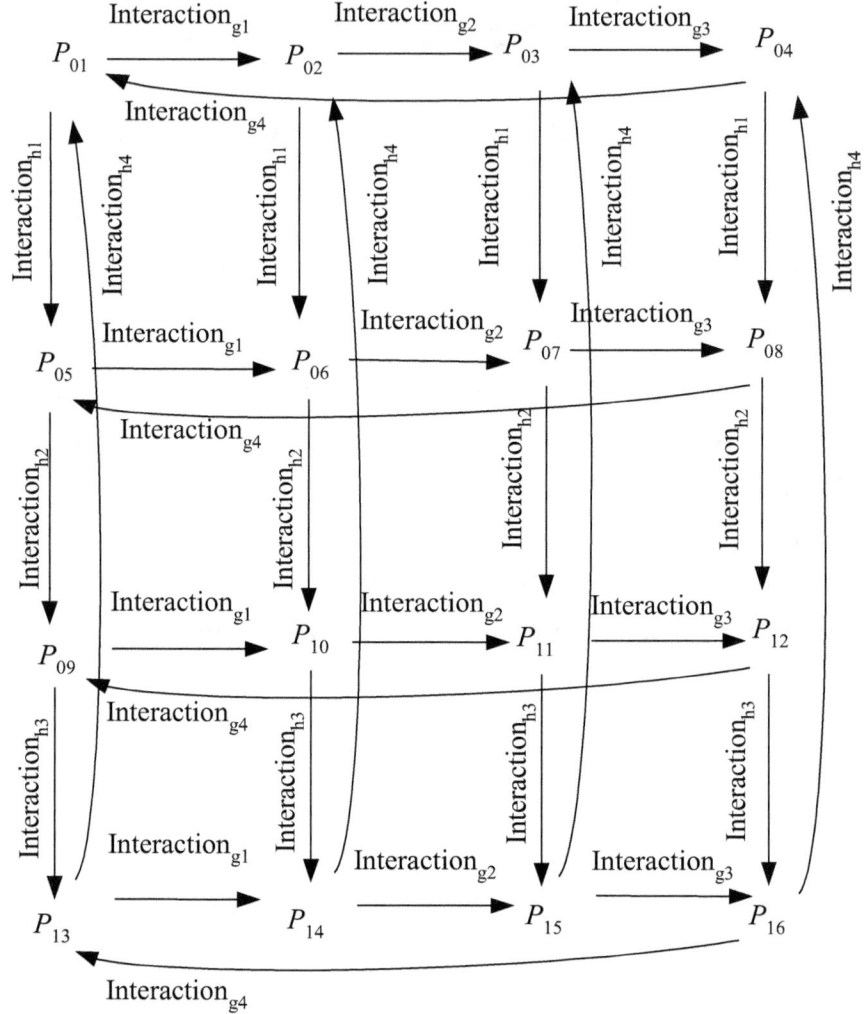

We syntactically represent multi-queue SBC process Q_{01} as $\text{FixIFD}_m \| \text{FixIFD}_n$ which equals to "**fix**$(X_{13}=$ $\text{Interaction}_{g1} \bullet \text{Interaction}_{g2} \bullet \text{Interaction}_{g3} \bullet \text{Interaction}_{g4} \bullet X_{13})$ $\|$ **fix**$(X_{14}=$ $\text{Interaction}_{h1} \bullet \text{Interaction}_{h2} \bullet \text{Interaction}_{h3} \bullet \text{Interaction}_{h4} \bullet X_{14})$". The following transition graph shows the semantics of process Q_{01}.

159

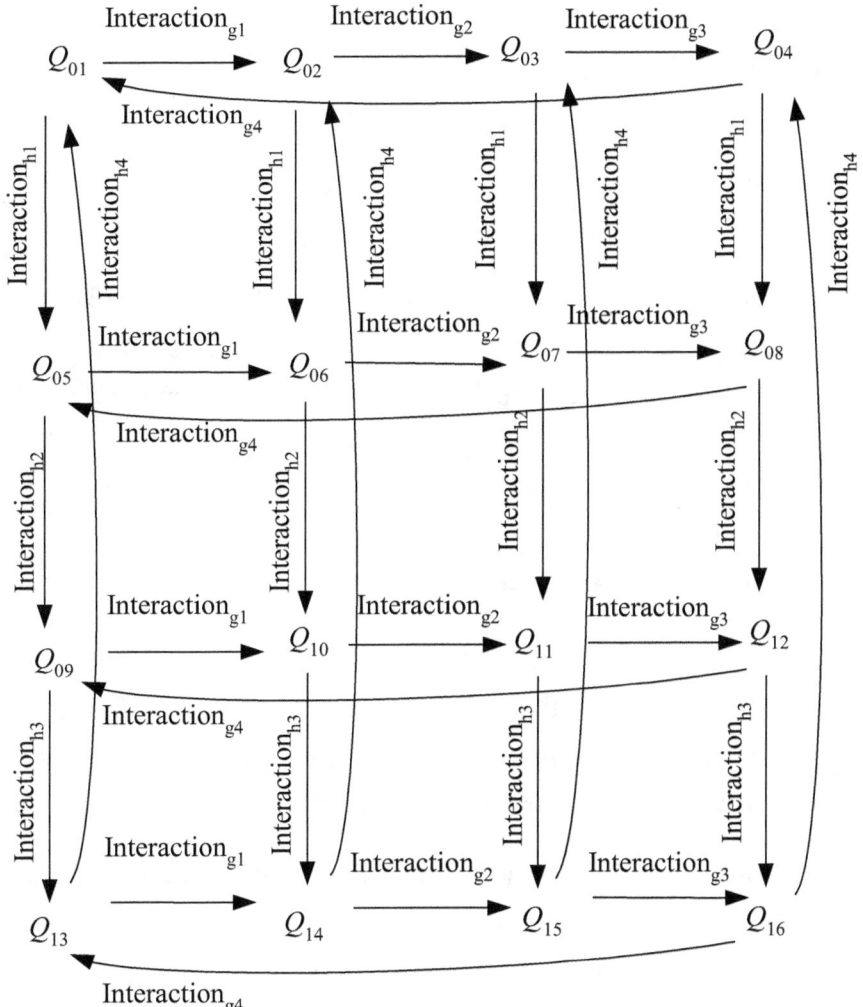

We can easily verify that $S = \{(P_{01}, Q_{01}), (P_{02}, Q_{02}), (P_{03}, Q_{03}), (P_{04}, Q_{04}), (P_{05}, Q_{05}), (P_{06}, Q_{06}), (P_{07}, Q_{07}), (P_{08}, Q_{08}), (P_{09}, Q_{09}), (P_{10}, Q_{10}), (P_{11}, Q_{11}), (P_{12}, Q_{12}), (P_{13}, Q_{13}), (P_{14}, Q_{14}), (P_{15}, Q_{15}), (P_{16}, Q_{16})\}$ is a bisimulation.

Using the S bisimulation, we then are able to verify that P_{01}

and Q_{01} are observation congruent because (1) P_{01} $\xrightarrow{\text{Interaction}_{g1}}$ P_{02}, then we have Q_{02} that Q_{01} $\xRightarrow{\text{Interaction}_{g1}}$ Q_{02} and $P_{02} \overset{\sim}{\approx} Q_{02}$, and (2) P_{01} $\xrightarrow{\text{Interaction}_{h1}}$ P_{05}, then we have Q_{05} that Q_{01} $\xRightarrow{\text{Interaction}_{h1}}$ Q_{05} and $P_{05} \overset{\sim}{\approx} Q_{05}$, and (3) Q_{01} $\xrightarrow{\text{Interaction}_{g1}}$ Q_{02}, then we have P_{02} that P_{01} $\xRightarrow{\text{Interaction}_{g1}}$ P_{02} and $P_{02} \overset{\sim}{\approx} Q_{02}$, and (4) Q_{01} $\xrightarrow{\text{Interaction}_{h1}}$ Q_{05}, then we have P_{05} that P_{01} $\xRightarrow{\text{Interaction}_{h1}}$ P_{05} and $P_{05} \overset{\sim}{\approx} Q_{05}$.

$P_{01} = Q_{01}$ means that "the design view of *Kurdi University*" and "the structural composition of the implementation view of *Kurdi University*" are observation congruent.

That is, we have demonstrated that the implementation view of *Kurdi University* is one level down structural decomposition (with observation congruence verification) of the design view of *Kurdi University*.

www.ingramcontent.com/pod-product-compliance
Lightning Source LLC
Chambersburg PA
CBHW051511170526
45166CB00001B/482